Tsunami and Fukushima Disaster: Design for Reconstruction

Rob Roggema · Wanglin Yan
Editors

Tsunami and Fukushima Disaster: Design for Reconstruction

Springer

Editors
Rob Roggema
Faculty of Design, Architecture and Building
University of Technology Sydney
Ultimo, NSW
Australia

Wanglin Yan
Faculty of Environment and Information Studies
Keio University
Fujisawa, Kanagawa
Japan

ISBN 978-3-319-85989-7 ISBN 978-3-319-56742-6 (eBook)
DOI 10.1007/978-3-319-56742-6

Printed on acid-free paper

This Springer imprint is published by Springer Nature
The registered company is Springer International Publishing AG
The registered company address is: Gewerbestrasse 11, 6330 Cham, Switzerland

Foreword

It gives me immense pleasure to write a foreword for this book. The book targets a very important and relevant topic of designing recovery. Six years have passed from the Great East Japan Earthquake and Tsunami of 2011. Different lessons are learnt in different parts of the affected areas, however, this book is possibly a unique attempt in English to describe the recovery process with specific and detailed field activities.

Community based recovery process has become popular in recent years, and there are different techniques used to receive community's views, perceptions and participation in the recovery process. This book exemplifies different design technology, which is used with the participation of local communities to get their views on the spatial issues of recovery. The book also illustrates the importance of blending local knowledge and external expertise to reach a workable and innovative recovery solution. The three key lessons in designing recovery area can be summarized as: flexibility and openness in local governance system, creating an environment and way to incorporate community's voice, and appreciating external expertise.

Within past six years, we have witnessed several other disasters in Japan, as well as in different parts of Asia and world. The 2013 typhoon Haiyan (Yolanda) of the Philippines has emphasized the need of risk perception and appropriate community actions. The 2016 Kumamoto earthquake has shown us the importance of lesson sharing within Japan itself. The lessons analyzed in the book is not only important within the affected region in Tohoku, but also can be applied widely in other parts of Japan. Moreover, the basic principles of community involvement in co-design recovery process can be used widely in the other disaster-hit areas in Asia and the world.

I sincerely believe that the findings of the book will be useful for the researchers, practitioners and decision makers, and this book will be a unique reference document for designing future recovery process. I congratulate the editors for bringing out this valuable piece of work, and to all contributing authors for their detailed analysis and important findings.

Prof. Rajib Shaw
Gradate School of Media and Governance,
Keio University, Shonan Fujisawa Campus (SFC)

Acknowledgements

The editors of this book would like to acknowledge the support for the design charrettes in Minamisoma and Kesennuma, which was received from the Australia-Japan Foundation (AJF). Besides this essential contribution, the workshops would not have been possible without the support of KEIO University in Japan.

Contents

Chapter 1
Introduction

Rob Roggema and Wanglin Yan

When in March 2011 the tsunami hit the east coast of Japan, all the disaster manuals were in place. Still the impact was devastating. Huge waves overthrew every coastal protection and caused the melting of the nuclear power plant in Fukushima. Within days the entire world understood the seriousness of the disaster. The manuals just couldn't cope. Many people died and the material damage was huge. The first learning the tsunami taught us is that not the manuals for the 'when' a disaster occurs will prevent the population from its impact. No, it is the planning and designing ahead of a disaster that might relief the impacts of a natural disaster. However, planning ahead of problems seems to be a most difficult task of planners and governments. They cannot respond with a plan if the problem is not concrete. The design must respond to the 'as if' of unforeseen, unprecedented events.

Even after a disaster takes place the design of the future proves to be difficult. The first attention goes, naturally, to victims and first aid. But after the first shock has ebbed away planners and governments can often not think of anything more than rebuilding the harmed settlements. This second stage of the disaster recovery is often undertaken without real involvement of the people living or returning to the area. Governments show leadership by acting swiftly with big, visible, rebuilding plans in order to satisfy the people. But are people satisfied in the long run? Or do they, in the end, oppose the measures that disrupt the, often subtle, environments

R. Roggema (✉)
Faculty of Design, Architecture and Building, University of Technology Sydney, Ultimo, NSW, Australia
e-mail: rob.roggema@uts.edu.au

W. Yan
Faculty of Environment and Information Studies, Keio University, Fujisawa, Kanagawa, Japan
e-mail: yan@sfc.keio.ac.jp

R. Roggema and W. Yan (eds.), *Tsunami and Fukushima Disaster: Design for Reconstruction*, DOI 10.1007/978-3-319-56742-6_1

people used to live in? For instance, the proposal of the Japanese federal government to implement enormous dams, most of them over 15 m in height, could count on strong resistance across the affected area.

In 2011, the Sendai region was suddenly confronted with multiple problems. Not only the direct impacts of the tsunami being flooding and radiation, but also secondary impacts were felt deeply in the area. People needed to leave their homes, some of them temporarily others permanently. Many people, especially farmers lost their jobs and could, once they could return, not be employed in their old jobs. Tertiary problems arose from this, such as the feeling of being useless and the loss of pride on the area where they were born.

In the lead off of the tsunami disaster in Japan the Sendai Framework for Disaster Risk Reduction was developed. The framework presents a guide for dealing with and preventing the risk of a disaster. When an area suffers from a disaster it can be used to alleviate the impacts and recover. The aim of the Sendai framework is to 'prevent new and reduce existing disaster risk through the implementation of integrated and inclusive economic, structural, legal, social, health, cultural, educational, environmental, technological, political and institutional measures that prevent and reduce hazard exposure and vulnerability to disaster, increase preparedness for response and recovery, and thus strengthen resilience' (UNISDR 2015).

The following priorities are set out in the framework:

1. Understanding of disaster risk in all its dimensions of vulnerability, capacity, exposure of persons and assets, hazard characteristics and the environment. This is a pledge for data collection, monitoring and assessing vulnerabilities and hazards;
2. Strengthening disaster risk governance for prevention, mitigation, preparedness, response, recovery and rehabilitation is needed to manage disaster risk. This consists of a clear vision, plans, competence, guidance and coordination within and across sectors, as well as participation of relevant stakeholders;
3. Investing in disaster risk reduction for resilience to enhance the economic, social, health and cultural resilience of persons, communities, countries and their assets, as well as the environment. These investments are also the drivers of innovation, growth and job creation;
4. Enhancing disaster preparedness for effective response, and to "Build Back Better" in recovery, rehabilitation and reconstruction.

On the basis of these priorities the Sendai framework identifies guiding principles (see text box). The focus of these principles is on inclusiveness and engagement of governments, stakeholders and 'all of society' in the process of recovery. In practice however the focus often is to technically repair the damage, such as the realisation of dams to create a safer situation. The role of citizens is underestimated, and is still not given full attention. If citizens are involved they are only brought in a position to respond to government propositions.

Guiding principles Sendai framework

- Primary responsibility of States to prevent and reduce disaster risk, including through cooperation
- Shared responsibility between central Government and national authorities, sectors and stakeholders as appropriate to national circumstances
- Protection of persons and their assets while promoting and protecting all human rights including the right to development
- Engagement from all of society
- Full engagement of all State institutions of an executive and legislative nature at national and local levels
- Empowerment of local authorities and communities through resources, incentives and decision-making responsibilities as appropriate
- Decision-making to be inclusive and risk-informed while using a multi-hazard approach
- Coherence of disaster risk reduction and sustainable development policies, plans, practices and mechanisms, across different sectors
- Accounting of local and specific characteristics of disaster risks when determining measures to reduce risk
- Addressing underlying risk factors cost-effectively through investment versus relying primarily on post-disaster response and recovery
- "Build Back Better" for preventing the creation of, and reducing existing, disaster risk
- The quality of global partnership and international cooperation to be effective, meaningful and strong
- Support from developed countries and partners to developing countries to be tailored according to needs and priorities as identified by them.

We want to introduce five stages of responding to a disaster:

1. *Recover*: In this first stage the concern is often immediate care, rescue and to provide temporary housing and amenities;
2. *Repair*: the second stage is characterized by the response: 'this should never happen again'. There is a strong push for making the protective systems stronger, such as the building of dams;
3. *Rebuild*: this third stage is concerned with building back what was before. By rebuilding similar urban environments in the same areas the same vulnerability is reintroduced. These areas may even become more vulnerable, as the risk at climate impacts increases and the population in these kinds or areas rises;
4. *Resettle*: this fourth stage is to resettle in the area, but not in similar urban environments that were there before. Resettling aims to increase the resilience, creating an urban system that better bounces back after impacted by a disaster;

5. *Redesign*: this fifth stage aims to redesign the area to anticipate a possible disaster and the urban environment is not vulnerable anymore. This requires a design of rethinking the landscape. The underlying landscape system with its water, ecological and soil systems form, together with the cultural history of the place the basis for urban design.

In every disaster area this five stages should be applied. Often the intentions of decision-makers, governments and planners are good, but in practice the attention for recovery of the area drops when the first 2–3 stages are over. We see this as a dangerous situation as it means that the area will be rebuild in very similar ways as it was before the disaster. Moreover, involvement of the local population in the planning process for future resilience is limited in these first phases. Only when stages 4–5 are deliberately and in all their broadness are executed, the transformation of the area, with involvement of the local population as co-creators in the planning process can be fully explored and implemented.

In this book two case studies are central. The redesigns of both Minamisoma as Kesennuma are extensively discussed, the content and even so the process. Both prefectures have been deeply hit by the disaster, each with their own typical impact.

In Minamisoma flooding of parts of the area forms a problem in combination with high radiation levels. This double complex problem has been troubling planners and decision-makers. Many initiatives, plans and policy documents have been written about Minamisoma. Many projects have been proposed and tried from public, private as well as citizen groups.

The main reason for redesigning the area of Minamisoma was the lack of pride the original inhabitants of the area had. A simple rebuilding process is therefore too simple. It undoubtedly would provide the new houses for people, and would even arrange a regenerated agricultural system, but the pride of the people would not return. The situation asked for rethinking the genes of the area and these were found in the cultural historic dimension and the strong relationship Minamisoma derives from the horse festival. This formed the key to let the area be reborn. Re-instalment of the horse festival, the horse keeping and extending the horse race with additional activities could regain the pride of the people, from which the people could start their lives and activities again. This made it possible for the people to return to their home ground. Therefore the motto of this redesign is 'Rethink-Reborn-Return'.

In Kesennuma the impact of the disaster is mainly an inundated area. This has caused replacement of people to higher grounds in temporary buildings, but it is uncertain whether people ever could return to the area of their washed away homes. There are numerous initiatives and plans made for Kesennuma.

The reason for the redesign of Kesennuma was the huge devastating impact of the flood as result of the tsunami. A huge wave flooded the area and inundated a large part. Many people died or were displaced. The lower area, close to the sea was washed away and the urban settlement that used to be there could not be rebuild. The linkages with historic culture were cut off, almost literally by the water. A simple relocation of people, providing them with permanent housing is therefore not enough. The people deserve their rebuilt town is embedded in its historic context. Only recognition of the disaster makes it possible to deal with it. Therefore the redesign firstly focused on remembering, not only what happened recently, but also what the stories reflecting history could tell. These aspects of recent and past history are relevant not only to remember them but also to reconnect the people with their environment. This mental reconnection is supplemented with concrete and physical connections, between the mountains and the sea, through the rivers and other infrastructure. This reconnection then made it possible to reform the area and design a future for a transformed coastal zone and a transformed hinterland. The motto for this redesign is therefore: 'Remember-Reconnect-Reform'.

The common issue in reconstruction of the devastated regions is how to accommodate the ideas and plans from various sources and bridge governmental departments, ndustrial sections and citizen groups to work together. Landscape planning and citizen participation in the design process form an effective platform to accommodate those initiatives and projects in a spatial way.

The applied approach in both design processes is the design charrette. This intensive way of designing together with inhabitants and stakeholders, with design experts and academics allows for local people to be involved in the design conception. This form of co-design holds the promise of integrated and supported designs. People feel attached to the outcomes and will support the implementation of the plans. Especially in disaster struck areas it gives people also the mental support to re-attach to their area, which gave them traumatic experiences before.

This books starts with chapter two in which Misato Uehara describes the tsunami disaster and the impact this had on the landscape in the region. This chapter is followed chapter three in which Rio Sakurai digs into the government activities and the importance of human recovery. In chapter four Roggema and Yan describe the design process applied for the redesigns in Minamisoma and Kesennuma. The design propositions for each of these regions described in detail in chapter five and six respectively. Chapter seven and eight contain a visual story of the two design processes in Minamisoma and Kesennuma respectively and chapter nine draws conclusions, makes recommendations and presents an outlook for planning and design in disaster areas.

Reference

UNISDR. 2015. *Sendai framework for disaster risk reduction 2015–2030*. Geneva: UNISDR.

Chapter 2
Post-3.11 Reconstruction, an Uneasy Mission

Wanglin Yan and Rob Roggema

2.1 Introduction

Six years have elapsed since the Great East Japan Earthquake struck the country's Tohoku region on 11 March 2011. A variety of projects are under way for the reconstruction of the disaster-affected area. The government declared the five-year period from 2011 until the end of March 2016 as the intensive reconstruction period, and is pouring considerable effort into the pillars of its reconstruction efforts, including support for survivors, restoration of public infrastructure, housing reconstruction and community development, reviving industry and livelihoods, revitalizing and reconstructing Fukushima and creation of a "New Tohoku" (http://www.reconstruction.go.jp/english/topics/Progress_to_date/pdf/201608_process_and_prospects.pdf). Over the course of five years, expenditures included 10 trillion yen for housing reconstruction and community development, 4.1 trillion yen for reviving industry and livelihoods, 2.1 trillion yen for health and living support for survivors, and 1.6 trillion for reconstruction and revitalization from the nuclear accident. The rebuilding of houses and reconstruction of towns and communities is being advanced through three approaches: independent reconstruction of housing, group relocation for disaster prevention, and disaster public housing improvement. As of January 2016, 49% of the approximately 30,000 units disaster-resilient public housing were completed, 32% of the approximately 20,400 units of new housing by independent reconstruction were completed, and

W. Yan (✉)
Faculty of Environment and Information Studies, Keio University, Fujisawa, Kanagawa, Japan
e-mail: yan@sfc.keio.ac.jp

R. Roggema
Faculty of Design, Architecture and Building, University of Technology Sydney, Ultimo, NSW, Australia
e-mail: rob.roggema@uts.edu.au

R. Roggema and W. Yan (eds.), *Tsunami and Fukushima Disaster: Design for Reconstruction*, DOI 10.1007/978-3-319-56742-6_2

restoration of 2308 schools was for 98% completed. Social infrastructure and seawall reconstruction was for 25% completed, national road reconstruction for 99%, and railway reconstruction for 93%. In industry, 74% of farmland could be planted, fishing port reconstruction was at 73%, and aquaculture facilities were 93% back in operation (Reconstruction Agency 2016).

These statistics show a steady progress in the reconstruction process. However, the delay in industrial and community recovery compared to public infrastructure progress is a concern. Even before the disaster, the region already had issues with a declining population, the aging of society, and the hollowing of industry. Through the reconstruction the government wanted to use Fukushima as a model for the revitalization of Japan. However, there is a large gap between ideal and reality. After five years, 182,000 people were still living in temporary housing. The population of 42 municipalities in the three Tohoku prefectures struck by the disaster had declined by an average of 10%, reaching the projected numbers of population levels for 2030 15 years early. Japan had seen a 140% increase in tourists nationally over five years, while the three prefectures had only reached 70% of their pre-disaster levels.

Also, the Fukushima nuclear accident left some serious challenges for reconstruction. A precondition for reconstruction is the clean-up of the nuclear accident, but there is no end in sight for removal of fuel rods, the final disposal of the reactors, and disposal of contaminated water, etc. A survey under citizens of Minamisoma City in September 2015 found that they still had profound concerns about the future, with 30% responding that they would like to live outside the district, city, and prefecture.

This disaster made people aware of the multiple threats of earthquakes, tsunamis and nuclear accidents. It is not possible to prevent a disaster in the context of the complexity of nature, although with proper approaches, it is possible to limit the extent of damage. Disaster countermeasures require an effort not only into disaster prevention but also disaster reduction. In other words, things should be done in a way that, even if a disaster occurs, recovery can be quickly, and society can recuperate as soon as possible. This is called the "resilience" approach. After the disaster, the government released its "Resilience Japan" concept and in 2012 passed the Basic Act for National Resilience Contributing to Preventing and Mitigating Disasters for Developing Resilience in the Lives of the Citizenry (www.japaneselawtranslation.go.jp). Based upon the legislation, the national, prefectural, and municipal governments basic plans for resilience have to be formulated. However, the legislation appears to emphasize disaster countermeasures for mega-quakes in the Nankai Trough and East Nankai Trough, and no special consideration has been given to reconstruction of the Tohoku region. For survivors of the Great East Japan Earthquake, their greatest hope is to escape as soon as possible from the post-disaster disruptions and to return to a new normal life. This restoration and recovery is another aspect of resilience.

Research regarding resilience began in the 1970s (Holling 1973), and today is seen as an essential factor for sustainable society (Zolli and Healy 2012). Resilience has two dimensions: resistance to disaster, and the ability to recover. Both of these are capacities that need to be developed for a country, region, organization, or system. In other words, first discover vulnerabilities for changes from the external environment, secondly build resistance, and then even if an incident occurs, the ability to absorb the disruption exists and limits the damage. After this adapt to environmental change, and transform to a better situation. When this is applied to disasters and reconstruction, this thinking means preparing well during normal times for the adaptation to disaster risk, absorbing the shock when a disaster occurs, then reconstruct quickly and make an effort to be stronger than before the disaster. This approach of absorbing, adapting, transforming, and thinking about change and transformation is called "resilience thinking" (Walker and Salt 2006).

A disaster is not desirable, but if the former system is damaged by a disaster, the subsequent reconstruction presents opportunities to build and increase resilience. However, restoration and recovery means taking action and they appear as a variety of short-term projects after a disaster. The capacity to recover is the capacity, which is built up from such actions, and it takes time to build this capacity. Finding a balance between the short-term and long-term is a major challenge in recovering from a disaster. Figure 2.1 summarizes these concepts.

In the figure, the vertical axis represents the living conditions (quality of life), the horizontal axis represents time, and the black dot shows the living conditions of survivors. If the disaster is severe, the decline in living conditions is significant. There are various barriers on the path to get out of that situation, and the approach to get there will depend on the vision for reconstruction. From this observation we can extract four key concepts:

1. The disaster itself,
2. The barriers to reconstruction,

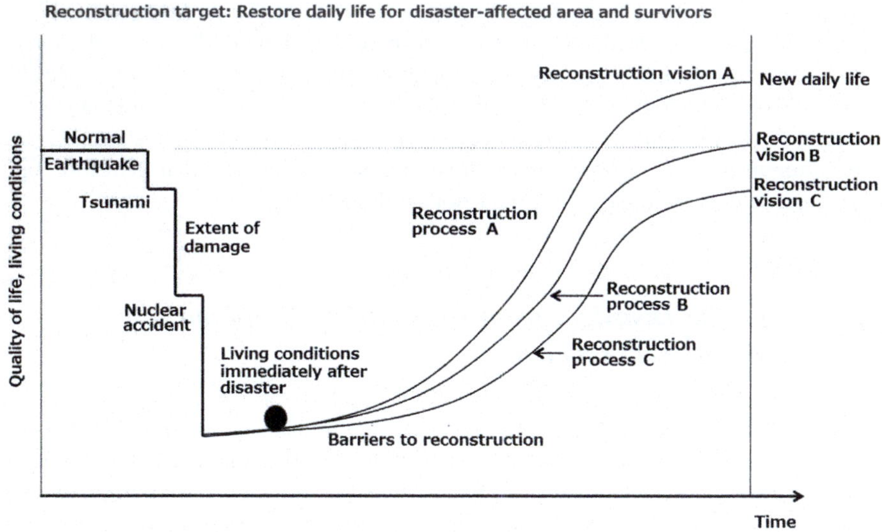

Fig. 2.1 Resilience approach to reconstruction (by author)

3. The reconstruction vision,
4. The reconstruction approach.

If the vision is clear, projects and recovery activities are easier to execute. However, the reality is complex. If the aim is to create better conditions than before the disaster (Reconstruction vision A), things will not go according to expectation unless the approach is commensurate. It is common to hear people say: 'We are making an effort but without results' and 'The idea is good but not permitted by the system'. In other words, reconstruction is not just a simple matter of technology. Japan is a mature society but did not establish clear answers to the following key questions: What is the nature of the disaster, what should be the aims of reconstruction, and what approaches are best?

With the aim of creating the locally-based capacity to recover, this chapter looks at the reconstruction from the perspective of resilience, presents approaches to promote co-creative reconstruction town planning, based on many stakeholders working together, and then considers the outcomes and challenges.

This chapter develops the concept of resilient reconstruction depicted in Fig. 2.1. Section 2.2 covers characteristics of the disaster, Sect. 2.3 covers barriers to reconstruction. This discussion lays a base for the co-creative reconstruction practice in later chapters.

2.2 Characteristics of the Great East Japan Earthquake

Some people say the reconstruction after the disaster was delayed, but perhaps the reconstruction plan was made with too much reference to the Great Hanshin Earthquake, also known as the Kobe Earthquake, which occurred January 17, 1995. In the face of the unprecedented disaster and unanticipated conditions of the Great East Japan Earthquake, some say the Japanese experience was ineffective. In the context of reconstruction and community planning, it is important to understand the character of the disaster. It is also important to recognise that the disaster conditions were different in Iwate, Miyagi and Fukushima prefectures, and that therefore, one single approach to reconstruction may not be the best.

2.2.1 Multiple Disasters from Earthquake, Tsunami and Nuclear Accident

The Great Hanshin Earthquake was an urban earthquake in a densely-populated region. An earthquake damages houses, buildings and structures, and there will be victims if they collapse. In the event of an earthquake, the local social infrastructure need to be rebuilt and lifestyles must be restored. The earthquake and tsunami of the Great East Japan Earthquake were of a size that occurs only once in a thousand years. Survivors of the tsunami needed to choose between moving away or returning to live in the original low-lying area. After the disaster, many of them wanted to move to higher ground to avoid any future tsunami. The government and researchers encouraged a reconstruction model, which separates employment and settlements, with employment being located in low-lying areas and housing on higher land. However, it is a long process, which takes time to find resettlement sites, develop plans, prepare the urban infrastructure, and construct housing. The Sanriku coast has only a finite amount of flat land, so finding resettlement sites is not an easy task. Also, to make flat land and low-lying areas safer, the ground level of the town must be raised. For the land to settle, it takes also time. Meanwhile, areas affected by the nuclear accident face another level of difficulty. The policies differ for a radius of 10, 20, and 30 km. In places where government functions have moved away, there are still no prospects for restoring them. Decontamination work is also not making progress, so it is still difficult to make any projections.

If the challenge was only to deal with earthquakes, considerable experience exists to build seismically robust designs for buildings and structures. However, as in this case of multiple disasters, normal assumptions may not be reliable. What is true safety, and what is practicality? It is attractive to live near the coast, but tsunamis are a major risk. Higher land is safer, but there may be not enough work. A town can be better protected if the seawall level is raised, but some people feel uncomfortable when they cannot view the sea. People thought nuclear power was a clean and convenient energy source, but it turned out not to be so. This disaster resulted in profound anxiety with residents, and it will take time and effort to alleviate that anxiety.

2.2.2 Damage to a Region with a Declining Population, an Aging Society, and a Weak Industrial Base

The six prefectures in the Tohoku region account for 17.7% of Japan's land area and 7.1% of the population, but it counts for only 6.0% of the gross domestic product, or GDP (METI 2011). In terms of industry, the Sanriku coastal region has fishery product processing and shipbuilding (fishing boats), but not enough manufacturing, knowledge, and service industries that attract young people. The residents here generally were living a self-sufficient life near the sea by cultivating oysters, scallops, ascidians (sea squirts) and *wakame* seaweed, while on land they grew rice. But this changed with industrialisation and urbanisation. After the period of rapid economic growth in the twentieth century, the population in the Sanriku region was steadily declining. Many towns were dealing a low birth rate and an aging population, and facing tough fiscal challenges. The disaster was a catastrophic shock, lifting fishing boats and dropping them inland, inundating fishing ports, and smashing seawalls. Much of the infrastructure has been rebuilt, including roads and fishing ports, but workers are not returning. Young people are less interested in fishing and farming. There are concerns that infrastructure, which was restored with big efforts may end up to be used very little. Reconstruction must not only mean restoring facilities, but a sustained socio-economic capacity should also be installed.

2.2.3 Disaster in a Time of Uncertain Economic Future

The disaster occurred after the global financial crisis of 2008, when the global economy was still unstable. The disaster interrupted supply chains, and many experts believed it would have a negative impact on the global economy. Meanwhile, some had the view that special demand driven by reconstruction would accelerate the economic recovery. The disaster resulted in enormous losses, as much as 16–22 trillion yen, but it was also seen as an opportunity, with the concentration of a considerable amount of money from financial government assistance, private sector casualty insurance, public donations, and so on. The past few years have even been referred to as a reconstruction bubble in the disaster-affected area. There is no clear social consensus, however, on where and how these enormous sums of reconstruction funds should be spent. Some have expressed the view that economic efficiency should be improved, by using reconstruction as an opportunity to select, concentrate, and place both industry and the labour force in large cities and industrial clusters. If this approach is chosen, it would mean abandoning fishing villages and small settlements. Certainly, concentration can boost efficiency. But when it comes to human beings, efficiency is not the only desirable factor to consider. Modernisation is based on the economic system. But the adverse effects of the earthquake disaster revealed weaknesses. Concentration depends on heavy and large-scale social infrastructure, and production, transport, and consumption creates enormous emissions of gases, wastewater, and waste, with negative impacts on the local and global environment and deleterious effects on sustainability of society as a whole. But sustainability of society does not mean prosperity for just one area of concentration or abundance for just one segment of the population. Also, due to its steep terrains, the Sanriku coast is perhaps not suited to be a society that prioritises efficiency. In this region, reconstruction using the conventional model that prioritises efficiency might not be the most sustainable one.

2.2.4 Natural Disaster or Man-Made Disaster?

This disaster was an earthquake that struck as an overwhelming external force, which triggered a tsunami, and resulted in the loss of a tremendous amount of life and property. This made it impossible to operate nuclear power plants, which could be considered a delicate system for modern society, resulting in enormous economic losses and social disruption. The earthquake and tsunami were unavoidable natural disasters, but the nuclear accident was not inevitable. Investigative reports and mass media portrayed this accident as human error and concluded that this was a human-caused disaster (Asahi Shimbun 2014).

A disaster is a natural and societal incident, and results in physical and human losses. We know that nature can be violent, and that humans can make mistakes. The potential to be exposed to an undesirable incident is known as risk. But even if that risk is known, it does not mean that one can immediately take evasive action. We make decisions in the midst of constraints on time, finances, strength, and capacity. Governments implement projects based on policy, and corporations conduct business based on finance. It is not possible to completely eliminate all risks in one sweep. It is not possible to prevent a disaster from occurring tomorrow. However, it is possible to imagine a variety of scenarios occurring in day-to-day life and to limit losses. That is the significance of resilience thinking. Disaster reduction requires consideration of disaster causes and effects, and integrated consideration of impacts on human activities and the behaviour of people in disasters. Losses that occur when these things are neglected could be all referred to as man-made or human-caused disaster. Seen from this perspective, this earthquake disaster was not only the nuclear accident, but it also included many elements of human-caused disasters in many dimensions. They are too numerous to mention, but some of these are: the building of primary and middle schools on low-lying land, the lack of evacuation roads, the failure of communications and notification systems to function during a crisis, and people should not have been living on low-lying land. Modern society is built upon fragile foundations like this, and this is at the root of human-caused disasters (Beck 1992). Reconstruction must re-launch itself from reflection about those things, and establish a resilient future oriented vision.

2.3 Reconstruction Challenges

2.3.1 Confused Visions for Reconstruction

Soon after the disaster, the national and local governments started to formulate reconstruction plans. At the end of 2011, the affected prefectures released their respective reconstruction plans. All of them portrayed a bright future with targets for livelihood restoration, industrial reconstruction, safety and security and liveability. After five years have passed, some stakeholders asked if the original plans were appropriate, and asked if they should be reconsidered (NHK 2015). The basic trend before the disaster was a declining population, but will the reconstruction process be capable of halting that trend? As suggested by Fig. 2.1, what should be the direction of reconstruction, and how far along the path can we call it reconstruction? The establishment of this vision is not easy. The resulting population and GDP might be lower than before the disaster. However, resilience is not just a matter of quantity, but can also be viewed as something that should also be measured in terms of human happiness and level of living conditions. The Reconstruction Agency regularly issues a report on reconstruction status and challenges, Iwate Prefecture uses a reconstruction INDEX, and Miyagi Prefecture issues reconstruction reports. However, they are still nothing more than lists of individual indicators, and have not reached the point of indicating the overall direction to be headed. We still tend to measure economic activity in terms of GDP. A steady-state economy is a sustainable economy that does not have GDP growth targets (Daly 1974; Yan and Tajima 2013). In contrast to “smart growth” there is also the idea of “smart retreat” (Science Council of Japan 2011). However, steady-state economics has never been tested for a declining population.

In its *Annual Report on the Japanese Economy and Public Finance 2012*, the year after the disaster, Japan’s Cabinet Office discussed the disaster and how the economic system should be (Cabinet Office 2012). It raised the idea of ‘the quality of development’ and asserted that households, corporations and society need to be more aware of ‘tail risks’ that have serious impacts but low probability of occurring, such as earthquakes and global financial crises, and that it is necessary to have resilient lifestyle foundations, corporate management, and a global economy. However, other than proposing the use of renewable energy, it made very few specific recommendations for broad and diverse reconstruction of disaster-affected areas. Nevertheless, it did ask questions that were not asked at the time of the Kobe Earthquake. Should reconstruction give a priority to efficiency in terms of the traditional economic system? Or should reconstruction emphasise the sustainability and quality of society and increase its resilience. Ultimately, resilience means responding with the capacity to spare, and that it is in itself incompatible with efficiency.

The 2011 disaster reminded people of the community and human connections and bonds in the Tohoku region. In the restoration as well, there is a tone of thinking about raising the level of happiness in terms of the local community (a given collection of people) rather than the individual (Tohoku Regional Advancement Centre 2012). However, an elevation of the level of happiness is something that is fostered over time. It cannot be synchronized with reconstruction targets that must be achieved in a short period of time. The question of how to close this gap is a major issue.

2.3.2 Decline of the Community

When it comes to risk, humans tend to have a strong interest in their surrounding area and what they can directly see. People will pay attention to family health insurance and employment insurance, but it is less common for people to take action proactively regarding the risks and future for the town as a whole. After the 2011 disaster, helping hands arrived from around the country, and connections spread out, and ‘Resilience Japan’ moved the world. But that was during a brief disaster utopia period (Solnit 2010). As the situation settled down, people became more distant from the activities, and communication declined. There was not necessarily a big opportunity for an inclusive discussion about the preferable direction of reconstruction. Also, discussions on "soft" (non-physical) dimensions, cultural, medium and long-term issues do not easily find their way to the government's "menu" of topics to address.

The Tohoku region was originally made up of hamlets and towns in an environment characterized by a ria coast, a coastline with multiple parallel inlets separated by prominent ridges extending inland, and with a high level of local self-sufficiency. A number of hamlets were combined to become towns, which the fishermen, craftsmen, and farmers shared. The individual hamlets had their own fire brigades, flood brigades, youth associations, chambers of commerce, housewives' groups, and so on, and communities were lively with activity. After Japan's post-war period of rapid economic growth, young people left, the local area shifted from primary industries to manufacturing and service industries, and organizations like this with local ties to the region dissolved.

Municipal governments affected by the disaster in Fukushima have been fragmented in a multi-layered way due to the tsunami and nuclear accident. The damage differed depending on the area. The conditions for reconstruction could differ on opposite sides of a road, for example in terms of tsunami flooding risk zone versus a housing restricted zone. So it was difficult for people could speak with one voice. In other places the issues were delayed in nuclear decontamination work, damage to brand reputation, or a decline in employment. Elsewhere, there could be concerns about future employment and health. It was not rare to see families be torn apart, for example, with the elderly members remaining, children being sent away to school, and parents going elsewhere to work (Zhang et al. 2014).

In this context, questions are being asked about what it means to emphasise the local community, and what will become of the community due to reconstruction.

2.3.3 *Project-Based Reconstruction*

Immediately after the disaster, the talks about the reconstruction raised expectations, as evidenced by slogans, such as 'Reconstruction is a re-launch from zero' and 'The future of Japan will start in Tohoku'. The year after the disaster, the government launched the Reconstruction Agency, in an effort to avoid the adverse effects of a vertical-splintered bureaucracy. However, the reality was that the budgets were formulated by the traditional government structure, with ministry jurisdictions. For example, for house reconstruction it was the Ministry of Land, Infrastructure, Transport and Tourism in charge, for decontamination the Ministry of the Environment, and for care of survivors, the Ministry of Health, Labour and Welfare. Even though reconstruction and community planning requires an integrated approach, the budgets were still isolated, on a project-by-project basis. As a result, gaps appear between projects (Akanuma 2014).

While terms like 'creative reconstruction' were used bandied about, each of the projects tended to follow conventional approaches and not take innovative steps. House reconstruction simply followed the standard government approach of public-managed disaster reconstruction, and only the minimum standards of disaster prevention design could be applied. The fishing port projects were nothing more than 'restoration' of damaged facilities, and coastal projects tended to be nothing but building 'seawalls'. Ideas that should have been considered, such as 'livelihood restoration', 'industrial recovery', and 'environmental harmony', did not apply. With these kinds of project-based budgets and project structure in municipal governments, the attention goes toward seeking how to acquire the budget and work proceeds in the usual way, without a complete picture of the future. For example, every town has a reconstruction plan, but in every one of them, the roads, facilities, and houses are planned simply in abstract terms. In many cases, when objectively reviewed later, people question whether that approach was good enough. If someone were to raise questions about how reconstruction could reflect a town's identity, the resources of the land, or incorporate peoples' wishes, it was rare to be seriously considered. Instead, the municipal governments, which should speak for the local people, were under the jurisdiction or authority of the national government or the prefecture, so they did not want to take responsibility. Unlike the model that propelled Japan into a developed country in post-war reforms, a model of creative reconstruction to put Japan on the path of sustainable development has not yet been found (Mikuriya 2016).

2.3.4 Roles of Government, Private Sector and Citizens

The disaster damaged natural and societal systems, so restoration has to be done based on new environmental conditions. For restoration and reconstruction after the disaster, different things are needed depending on the phase. To respond to this situation, new information and methods are needed. When people are asked what changed in Tohoku with the disaster, many will say that it was the flow of people and information. After the disaster, civic participation was active in every town. Non-profit organisations and university students flowed into the region, and experienced the disaster together with the people, spoke about hopes for reconstruction and prepared plans. The government, as well, paid attention not only to hardware, the physical construction, but also to software. About 20% of the reconstruction funds for 'New Tohoku' were allocated to non-physical projects. The majority of those funds were allocated to reconstruction projects done by citizens and NPO (Non-Profit Organisation) stakeholders. The projects covered a broad ranges of areas, such as renewable energy, care for people living in temporary housing and tourism promotion. In response, many small and medium-sized organisations started, including NPO's and general incorporated associations, a type of government designation for registered organisations. Immediately after the disaster, private sector corporations were also highly visible and companies around the country showed their desire to participate in reconstruction projects.

However, it should be noted that the expectations of corporations, NPO's and universities differ from each other, so it is not so simple for all to come together in a unified way for a creative reconstruction. After the disaster, a wave of helpers and participants rushed into the area and in some cases the locals were overwhelmed. The external NPO's, universities and corporations did not adequately grasp the local needs, and had a tendency to impose their own ideas. In that situation, in Iwate Prefecture, Iwate University played a central role and the government created a liaison offices and contact points. In Kesennuma City (Miyagi Prefecture) as well and universities played an active role. Forty universities got involved, created a network and supported the reconstruction. NPO's and general associations also played a role in providing intermediary support. However, as activities shifted toward reconstruction, the needs became more diverse, making it more difficult for high quality assistance to be provided by people serving as volunteers. At this stage of reconstruction, it is important to have frameworks in place that can coordinate internal and external collaboration.

2.4 Conclusions

The Great East Japan Earthquake of 2011 was a multiple disaster that occurred in the midst of a complex domestic and international situation. For reconstruction it is necessary to provide urgent assistance so that survivors can quickly free themselves from the conditions of the disaster, and also to balance those needs with a future plan that considers sustainability. As approaches for that, we often see government-directed citizen-participation models, but in the race against time, reconstruction often does not advance the way people may have expected. Meanwhile, with the reconstruction of the 2011 disaster, many experts went into the local area and cooperating on reconstruction with diverse stakeholders, including universities, governments, citizens, and companies. In such a context, the government, citizens, and experts, thought together about the future, without distinction or hierarchy, and from that situation the potential was initiated for plans and design. This chapter refers to that approach as 'co-creative reconstruction and community planning'. In this chapter the authors describe their experiences of having participated in such processes in Minamisoma City and Kesennuma City, and explain the details of these approaches.

References

Akanuma. 2014. Four barriers, Tokeiren, Aug 2014 (in Japanese).

Asahi Shimbun. 2014. Yoshida report. http://www.asahi.com/special/yoshida_report/. Last access date 24 Apr 2016 (in Japanese).

Beck, U. 1992. *Risk society: Towards a new mordernity*. London, Thousand Oaks, New Delhi: Sage Publications. Retrieved from http://www.amazon.com/Risk-Society-Modernity-Published-association/dp/0803983468.

Cabinet Office, Government of Japan. 2012. http://www5.cao.go.jp/j-j/wp/wp-je12/h02_03.html. Last accessed 24 Jan 2017 (in Japanese).

Daly, H. 1974. The economics of the steady state. *The American Economic Review* 64 (2): 15–21.

Holling, C.S. 1973. Resilience and stability. *Annual Review of Ecological Systems* 4: 1–23.

METI. 2011. Tohoku Today by Data. http://www.tohoku.meti.go.jp/kids/databook/. Last access date 24 Apr 2016 (in Japanese).

Mikuriya. 2016. Thinking about Japan from Post-war to Post-disaster, Nihonkeizaisinbun, 10 Mar 2016. http://www.nikkei.com/article/DGXMZO97981080T00C16A3000000/. Last access date 24 Apr 2016 (in Japanese).

NHK. 2015. What's needed in the 5th year of reconstruction, Kaisetsu Studio, 21 Mar 2015.

Reconstruction Agency. 2016. The Status quo of reconstruction and prospects, Mar 2016. http://www.reconstruction.go.jp/topics/main-cat1/sub-cat1-1/160315_mitinoritomitoshi.pdf. Last access date 24 Apr 2016 (in Japanese).

Science Council of Japan. 2011. Strategies for sustainable revitalisation of the nation and regions. http://www.scj.go.jp/ja/info/kohyo/pdf/kohyo-21-t133-1.pdf. Last access date 24 Apr 2016 (in Japanese).

Solnit, R. 2010. *A paradise built in hell: The extraordinary communities that arise in disaster*. Penguin. Retrieved from http://www.amazon.co.jp/Paradise-Built-Hell-Extraordinary-Communities-ebook/dp/B003XQEVLM/ref=sr_1_4?ie=UTF8&qid=1447652705&sr=8-4&keywords=rebecca+solnit.

Tohoku Regional Advancement Centre. 2012. Report on the qualification of Happiness. http://www.kasseiken.jp/pdf/library/guide/24fy-0601.pdf. Last access date 24 Apr 2016 (in Japanese).

Walker, B., and D. Salt. 2006. *Resilience thinking: Sustaining ecosystems and people in a changing world*. Island Press.

Yan, W., and H. Tajima. 2013. *Toward the Sustainable Development of Asi: In perspective of Economy, Society and Culture*, p. 340. Keio University Press.

Zhang, H., W. Yan, A. Oba, and W. Zhang. 2014. Radiation-driven migration: The case of Minamisoma City, Fukushima, Japan, after the Fukushima Nuclear Accident. *International Journal of Environmental Research and Public Health* 11 (9): 9286–9305. doi:10.3390/ijerph110909286.

Zolli, N., and A.M. Healy. 2012. *Resilience: Why things bounce back*. Headline Review.

Chapter 3
The Lessons Derived from 2011 Tohoku Earthquake and the Repercussion of the Myopic Decision-Making Structures

Misato Uehara and Wanglin Yan

Abstract The holistic land use planning process might contribute to reduce the risk of natural disaster and mitigate climate change. This also reduces huge national budget waste for reconstruction and social care after disasters. The aim of this study was to reveal the lessons derived from 2011 Tohoku Earthquake, Tsunami, and the Fukushima nuclear accident. The history of 2011 Tohoku disaster area's land use planning and actual land use change was analyzed. This found that destroyed developments by 2011 disaster have an essentially similar issue; bad repercussions of the myopic decision making structures. It suggests that holistic decision making with interactive land use planning process is important.

3.1 Introduction

The magnitude 9.0 Tohoku Earthquake occurred at 14:46 on 11 March 2011, in front of the east coast of Japan. The pacific plate sub-duction zone caused this earthquake and the huge (two to twenty meter high) tsunami. The destructive aftermath of this natural disaster lead to an irreparable situation. Not only for Japan, but also for the whole world were hazardous and caused an unexpected tragedy because the damages of the nuclear power plants.

The human damage by this disaster was 15,889 casualties and 2594 missing (as of 9 July 2015). More than 400,000 buildings were demolished or partially destroyed.

Later, it was discovered that the seaside area sunk one meter and the land of Tohoku was pulled 5.3 m into the Pacific Ocean. Moreover, the explosion and

M. Uehara (✉)
Graduate School of Science and Technology Ina Campus Landscape Planning Laboratory, Shinshu University, Nagano, Japan
e-mail: ueharam@shinshu-u.ac.jp

W. Yan
Faculty of Environment and Information Studies, Keio University, Fujisawa, Kanagawa, Japan
e-mail: yan@sfc.keio.ac.jp

R. Roggema and W. Yan (eds.), *Tsunami and Fukushima Disaster: Design for Reconstruction*, DOI 10.1007/978-3-319-56742-6_3

demolition of the Fukushima Daiichi Nuclear Power Plant caused radioactive contamination in an area around the power plant, with irreversible damages to the environment.

This nuclear accident immediately followed the earthquake and tsunami disaster. 400,000 people had to leave their hometown to another, safer place as result of this triple disaster. The International Nuclear Event Scale added this nuclear power plant accident to the severest level (rank 7) of nuclear accidents, similar to the Chernobyl disaster in 1986.

Most of the undertaken research of the East Japan great earthquake is aiming to discuss facts. For example, Fujii et al. (2011) and Maeda et al. (2011) reported a tsunami initiated waveform inversion by ocean-bottom pressure and GPS wave, which can observe the tsunami sea water level change. Mori et al. (2011) presented a regional dependence of tsunami characteristics. Some studies consider the regional planning issue. Kanako et al. tried to compare situations of the local reconstruction planning of Tohoku disaster area in their first year. Matanle (2013) also indicated a mismatch between the reconstruction planning and regional situation of a shrinking population and community aging. Zare (2012) framed the Fukushima Daiichi nuclear plant's accident as administrative crisis management in his analysis. And finally, Takeuchi et al. (2014) proposed the conversion of regional planning from engineering to ecological resilience.

However, few researchers have addressed the problem of overall consideration of why was this disaster so cruel. In fact, the analysis of man-made disaster (spatial planning failure) seems insufficiency while this disaster has both natural and social aspects. This chapter tries to focus on some of the lessons learnt from the man-made failure of the disaster.

This analysis brings us two important notes from the viewpoint of regional planning. The first note is that decision-maker should consider not only development area, but also surrounding area; in many cases, development area seems benefits recipient, surrounding area seems potential disadvantage recipient. The second note is that the benefits and costs of regional development should be considered before decision-making.

3.2 What Happened in Fukushima and the Tohoku 2011 Disaster Region, Before and Short After the Disaster

Figure 3.1 shows differences of early information about radioactive contamination from Fukushima Daiichi nuclear plant provided by Japan and France. Unfortunately, Japanese radioactive contamination prediction depended on only linear distance from Fukushima Daiichi nuclear plant: Japanese government offered evacuation instruction to residents who live within 20 km from Fukushima Daiichi nuclear plant. They also offered sheltering indoors instruction to residents who live within 30 km (Fig. 3.1 left).

However, the actual radioactive contamination crossed over the divided administrative boundaries by the local wind blowing at the time of the disaster, as the France simulation, IRSN (2011) shows (Fig. 3.1 right).

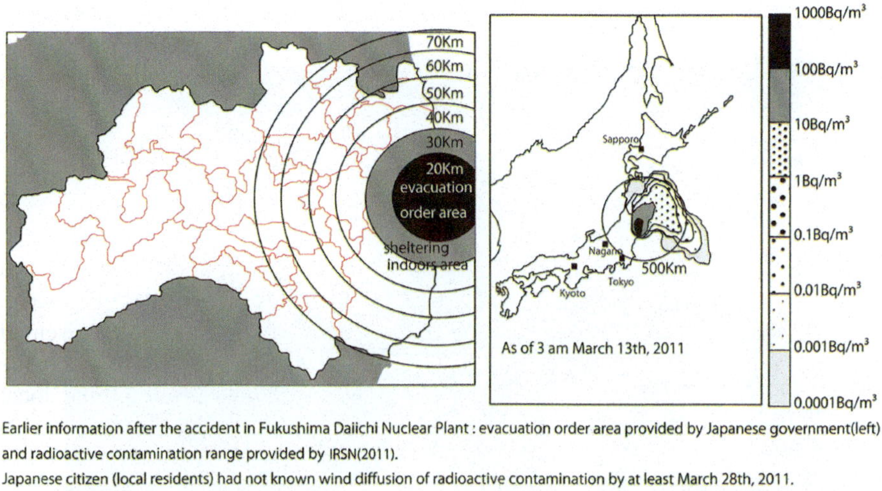

Fig. 3.1 Difference of Japanese and French early warning after the accident in Fukushima Daiichi nuclear plant

Fukushima's nuclear radioactive diffusion teaches us that the potential disadvantage recipient area is not corresponding with the power plant developed site. In the 1970s Fukushima and Japanese Government decision-makers, and nuclear experts did not consider the possibility of the potential disaster disadvantage outside the nuclear plant-building site. In 1980–1990, there were some assignments to review the potentially risk of the nuclear plant-building planned site by landscape architects and another public sectors: National Land Agency of Japan.

However, from 1980 to 2010 decision-makers and nuclear experts did not deal with the possibility of the potential disaster disadvantage recipient (area) outside of the benefit recipient either, because this additional readjustment required a lot of extra money and time.

Figure 3.2 shows the damages caused by the tsunami in Taro-cho, Iwate Prefecture, which had the largest seawall in Japan. These seawalls were built in all the coastal areas in Tohoku area and defended the land from an assumed small tsunami, in order to make possible for many people to live in the coastal area. This seawall gave the Tohoku coastal region an advantage compared to the other, less protected, coastal regions.

However, houses build in the coastal area, which was protected by the huge seawall was destroyed by the tsunami on March 11 (Fig. 3.3). Some people did not notice the tsunami disaster because of the huge seawall, and some were killed by the concrete block of broken seawall same as a lot of Tohoku coastal municipalities. When a natural disaster exceeds supposition of tsunami level, same problems will occur, though the reconstruction of a higher seawall is now advancing.

Indeed, the developer can sell a numbers of these building estates in short period, on reclaimed easy to occupy flat parts of the coast. It seems that there was a hefty

Fig. 3.2 10 m high seawall and coastal town destroyed by 2011 tsunami

Fig. 3.3 Former tsunami disaster area was destroyed again by 2011 tsunami, despite they had the largest seawall in Japan

profit taken by a particular kind of party. However, both the benefit and the costs of regional development for many involved parties should be considered before decision-making. In this case, the short-sighted benefit is the ease of land acquisition for the developer, while the less-visible costs of residents and the government were the land vulnerability for earthquakes and tsunami disaster. It is necessary to consider both elements at the same time.

3.3 What Happened in Fukushima and Tohoku Region After 2011 Disaster

In addition to what we discussed above, we would like to think about a new problem in the tsunami stricken area after 2011 disaster. Figure 3.4 shows a landscape change simulation of a 10 m seawall development plan in a beautiful fishing village (Karakuwa-cho, Shibitachi-chiku) in Kesennuma City, Miyagi Prefecture. Because of a complicated coastal landform (rias coast) in this area, an extraordinary high seawall is planned in this area after the Tohoku earthquake and tsunami disaster. This visualization shows that the environment of continental areas and the sea is completely divided. The seashore environment will disappear completely because of the width and height of the new seawall has to be bigger than existed design standards. Indeed, old seawall basement was destroyed by 2001 tsunami, and these seawalls could not save about 15,000 coastal people lives.

Fig. 3.4 Landscape change prediction of plan for a 10 m seawall after 2011 disaster in a beautiful fishing village (Miyagi prefecture). After this simulation, a part of seawall in this area was stopped by Miyagi-prefecture, for resident's protest

The planned 10 m-seawall might protect people by a 2011 scale tsunami, which is anticipated to occur once every 1000 years. However, this huge seawall brings a conflict between residents in the disaster region, which is comparable to the nuclear development's conflict as follow. This plan causes potential disadvantages for fishery and tourist business, and this is one of the reasons why the population in these coastal areas is decreasing after the 2011 disaster.

3.4 Lesson Learnt from This Disaster for Man-made Failure

Next, we consider what can be learnt from this East Japan great earthquakes area's regional development issue. We think the issues of the three-area developments described above (cases with each a different environmental development) have essentially similar issues (Table 3.1).

The first common issue seems that development decisions are made by an extremely limited number of stakeholders. In other words, it seems that the impact of the disaster in 2011 was aligned with the opinions of the organization and people in the region who get the development profit directly.

For instance, the Fukushima Daiichi nuclear plant, which had an accident in March, 11th, 2011 is constructed on the boundary of Futaba-machi and Okuma-cho of Fukushima region. However, radioactive contamination of Fukushima daiichi nuclear plant expanded way beyond the area of Futaba-machi and Okuma-cho. Moreover, the whole nation had subrogated the loss caused by the accident of the nuclear plant, using national tax for accident response beyond Fukushima region.

Figure 3.5 shows the boundary of Fukushima's coastal municipalities and the radioactive contaminated terrain resulting from 2011 nuclear accident. This figure shows the radioactive contamination in Namie-cho and Iitate-mura where more serious contamination is found than Futaba-machi and Ookuma-cho, which actually had a nuclear power plant in their area. The Namie-cho neighbourhood protest movement for nuclear plant was the only example of prevention for nuclear plant development by Tohoku Electric Power Co., Inc. Moreover, Iitate-mura was advanced village for sustainable regional planning in a co-operation with professor Itonaga Koji of Nihon University. Professor Itonaga introduced the Centre for Alternative Technology (C.A.T.)'s eco-design and Permaculture to Japan. In November 2014, Iitate-mura's 3000 residents demanded additional indemnification from the Tokyo Electric Power Company (TEPCO).

The second common issue seems the lack of development disadvantage simulation. Figure 3.6 shows money income and outgo of regions in Fukushima by Fukushima Daiichi nuclear plant's enticement. We made Fig. 3.6 from many separate data of income and outgo of Fukushima Daiichi nuclear plant's enticement; Agency for Natural Resources and Energy (2011), All Japan Council of Local Governments with Atomic Power Stations (1998) and NHK NEWS Web (2014).

Table 3.1 Three different developments in Tohoku disaster area have an essentially similar issue

Period	Development object	Development scale	Promoter	Opponent	Overlooked disadvantage	People who owe disadvantage without development decision making
Before the accident	Nuclear power plant	Large	• Assembly of Futaba-machi and Okuma-cho • Japanese government • Nuclear power exparts	Part of local community	• Radioactive contamination • Loss of local industry and population	• Surrounding manucipalities • Japanese government • Surrounding nations
	Coastal new town	Middle	Land owner of development site public works ContTantor	Nature Conservation Group	• Residence Liquefaction by Earthquake • Tsunami Damage • Disaster Refuse Disposal	• New Residents who thought the site safe • Generation who succeeded to a house which suffered from Totidoj disasters
March 11th, 2011						
After the accident	Seawall	Small	• Land owner of development site • Public works contrantor	Part of local community (Mainly fishries industry and tourism industry)	• Fishries industry reduction • Tourist decrease by landscape change • Depopulation and population aging	Next generations who have to pay for its developments and maintenance

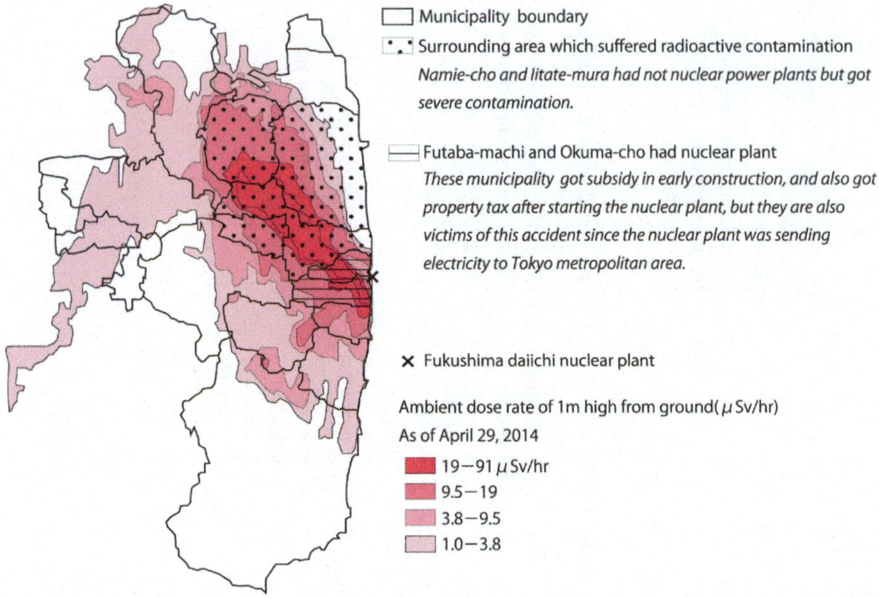

Fig. 3.5 Unequal pollution by radioactive diffusion across Fukushima's coastal municipalities, created based upon data from Ministry of Education, Culture, Sports, Science and Technology (2011)

This analysis teaches us the imbalance of Fukushima region's money income and outgo by Fukushima Daiichi nuclear plant's enticement: the money outgo after 2011 disaster (only 4 years) is bigger than the income in the 40 years before. In addition, Table 3.2 shows the breakdown of money income and outgo of region in Fukushima by Fukushima Daiichi nuclear plant's enticement. In particular, we have to recognize that the 2011 Fukushima Daiichi's accident have given surrounding area serious money outgo by the radioactive contamination. Despite this, the money income during 40 years became the benefit of a narrow area where the Fukushima Daiichis nuclear plant was built (Fig. 3.6, Table 3.2).

There are regions, which have income and outgo as Fukushima Daiichi nuclear plant's enticement and accident. This imbalance, what we call human-made disaster was as tragedy as 2011 natural earthquake and tsunami disaster. In other words, basically Fukushima Daiichi nuclear plant's enticement was promoted by limited sector, which had received the benefit.

Moreover, past new town development on the coastal area before 2011 disaster (mostly economic growth period from 1970 to 1990) and present huge seawall planning also had been promoted by limited stakeholders which had received the benefit: limited status appears land owners and civil engineering and construction's engaged person, in this case.

As mentioned above, we found that these series of large-scale development works were executed for limited numbers of people who obtain the profit, although

Table 3.2 Breakdown of money income and outgo of region in Fukushima by Fukushima Daiichi nuclear plant's enticement

		Before the accident					After the accident	Total
		1961–1970	1971–1980	1981–1990	1991–2000	2001–2010	2011–2014	
Income	Futaba-machi Okuma-machi	4.6	4.2	2.3	2.0	2.0	0.0	75.5
	Fukushima prefecture	0.8	0.7	0.7	0.7	0.7	56.9	
Outgo	Futaba-machi Okuma-machi	–	–	–	–	–	−140.9	−1103.1
	Fukushima prefecture and each municipalities (except Okuma-machi and Futaba-machi)	–	–	–	–	–	−359.1	
	Outside of Fukushima prefecture	–	–	–	–	–	−603.1	

100 million dollar = 10 billion yen

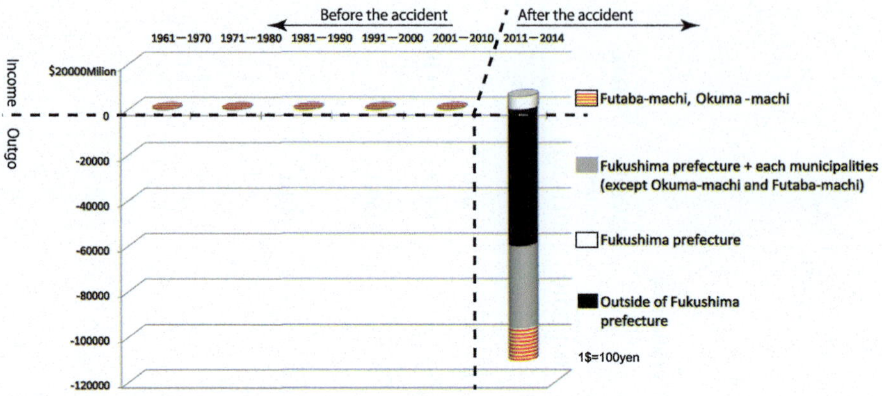

Fig. 3.6 Money income and outgo of region in Fukushima by Fukushima Daiichi nuclear plant's enticement

there was a small protest campaign during these developments. The influence of disadvantage of developing site and around area was disregarded at the time.

Consequently, it seems that these unequal development decision processes related to the above-mentioned area was not improved before or after the East Japan great earthquake. In other words, the Tohoku disaster areas repeated similar failures in different regional developments over times, excluding very few success, which set reconstruction work holistic landscape planning (Uehara et al. 2015).

3.5 Overview of Land-Use Change by Comparative Three-Dimensional Photographic Analysis

In this section, we try to compare the 3D photographs for reviewing tsunami disaster area development. We made these 3D photographic images using time series of aerial photos of the Geospatial Information Authority of Japan by Agisoft.

Figure 3.7 shows us coastal concave landform of the Fukushima Daiichi nuclear plant site.

We can find past salt farm that located at coastal concave landform. This coastal concave became a site of Fukushima Daiichi nuclear plant. After 2011 disaster, the Fukushima Daiichi nuclear plant site has seriously contaminated water problem, for nuclear cooling and underground water flowing. We find one of the reasons of contaminated water problem seems coastal concave landform. The Tokyo Electric Power Company (TEPCO) has tried to block underground water flow into the Fukushima Daiichi by underground ice wall (Hyodo-heki) construction. This high-tech engineering has not solved this increasing polluted water problem, although huge national budget (13 billion US dollars) was used.

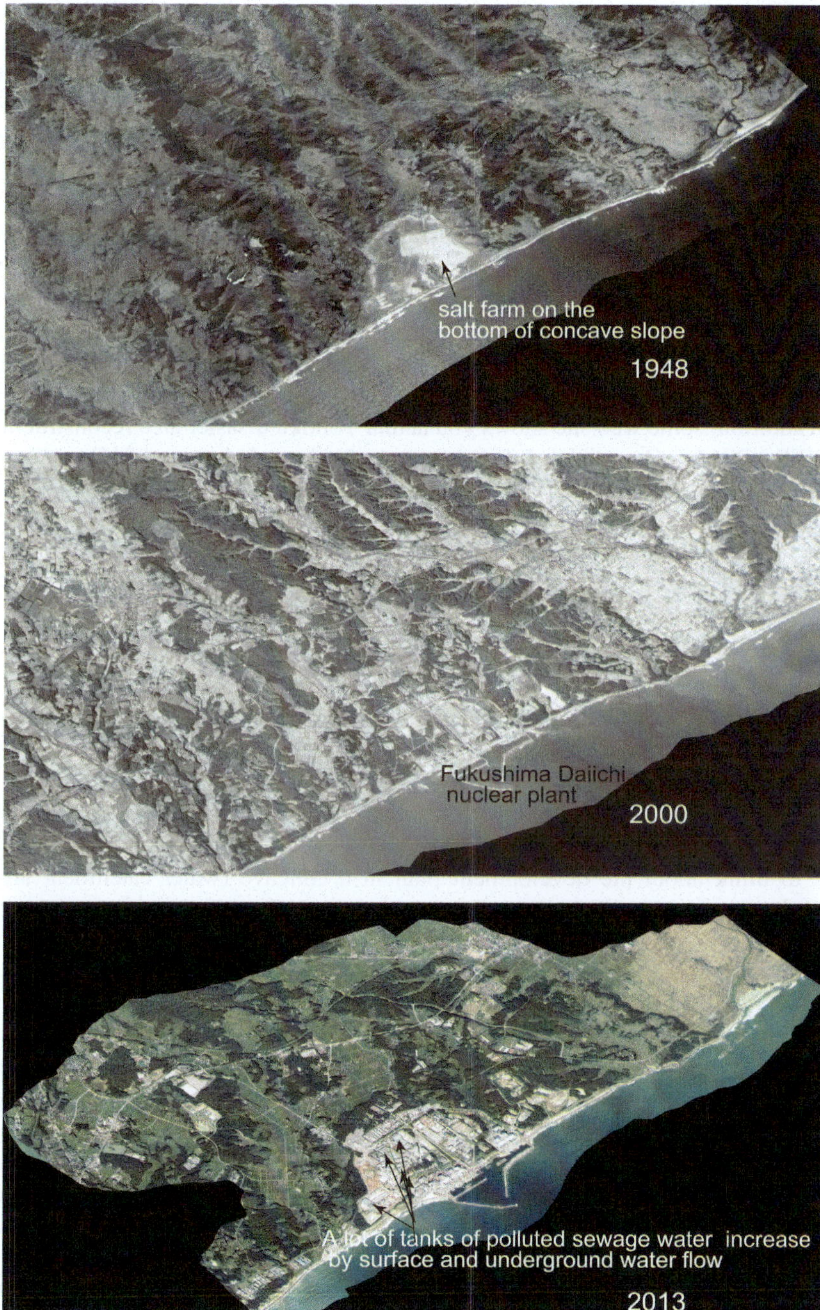

Fig. 3.7 Comparison of the before and after of 2011 disaster at the Fukushima Daiichi nuclear plant site's land use, created based upon data from Geospatial Information Authority of Japan (2015a)

Figure 3.8 shows Taro-cho district's land use transition from 1948 to 2013. We find beautiful ria-coast land use in 1948. In 1948, most part of Taro cho's bay was natural open space: sand beach and grassland. Until 2000, these coastal open space became new town, which were protected by X shape layout seawall. In 2011 March 11, this deeply indented coastline made tsunami level higher than flat coastal area's tsunami. In addition, X shape layout seawalls turned tsunami direction into seawall node point, so tsunami destructibility became strong.

After 2011 disaster, Taro-cho has reconstructed two separate seawalls on same area. However, these seawalls layout improvement area remain difficult to evacuate because big 10 m seawalls hinder vision (Fig. 3.9).

These two overviews of land use change direction teaches us the land use planning, which adapts local environment is more important than individual high-tech engineering solution.

In addition, some web site shows you actual Tohoku disaster area reconstruction; NAVER (2015), Geospatial Information Authority of Japan (2015b), Google (2015) and Dailynewsagency (2015).

3.6 What Can We Do in the Future for Sustainable Development?

Finally, we would like to consider what could be done in the future. Hereto, in the field of landscape planning, the overall development planning method has been researched.

McHarg (1969) method is a basic overall planning process model, which points us to consider the development influence area as region setting. For example, we have to think about the development plan in a concave (water catchment area), which has special interactions as a unit with other areas. The underground water flow and the monsoons wind make environment to one ecological region. In addition, we have to consider disaster vulnerability of the planning site, (Uehara 2012) as a case study of Tohoku disaster prediction issue.

To our regret, the site selection of the nuclear plant in Fukushima did not consider the relativity of the above-mentioned environment. A former salt farm site became a site for Fukushima Daiichi nuclear plant because of a simple reason: there is abundant seawater available for nuclear reactor cooling, and the land acquisition was easy. But, the level of polluted water increases at the pace of 400 t/day in the Fukushima Daiichi nuclear power plant, because underground water flows into the contaminated site every day as result of the coastal ebb and flood tides, and concave landforms.

We analysed relationship between the boundary of Tohoku prefecture, the location of nuclear plants, and the prevailing wind map of National Land Agency Japan 1980s ecological planning database (Fig. 3.10). This relationship of prevailing wind, prefecture boundary, and location of nuclear plants teach us the

Fig. 3.8 Comparison of the before and after of 2011 disaster at the Taro-cho's district land use, created based upon data from Geospatial Information Authority of Japan (2015a)

Fig. 3.9 Before and after the 2011 disaster at the Taro-cho district landscape. Although the reconstruction advances, the town structure is kept, which goes against the tsunami disaster with huge seawall

possibility of radioactive contamination reaches exceeding the range of each municipality.

This holistic landscape planning aspect can become not only for the protesting campaign a reason for development, but also functions as positive response material for evacuation when an accident actually happens.

Moreover, the residents' participation in the planning process (Roggema 2012a) is also extremely important, because limited stakeholder's discussion and decision making often exclude alternative discussion completely. For instance, in the past the promotion sector believed nuclear technology is absolutely safe, reason why they neglected disaster measures. In these planning processes, discussion tools can be introduced to offer participants the room for entering their opinions, such as the

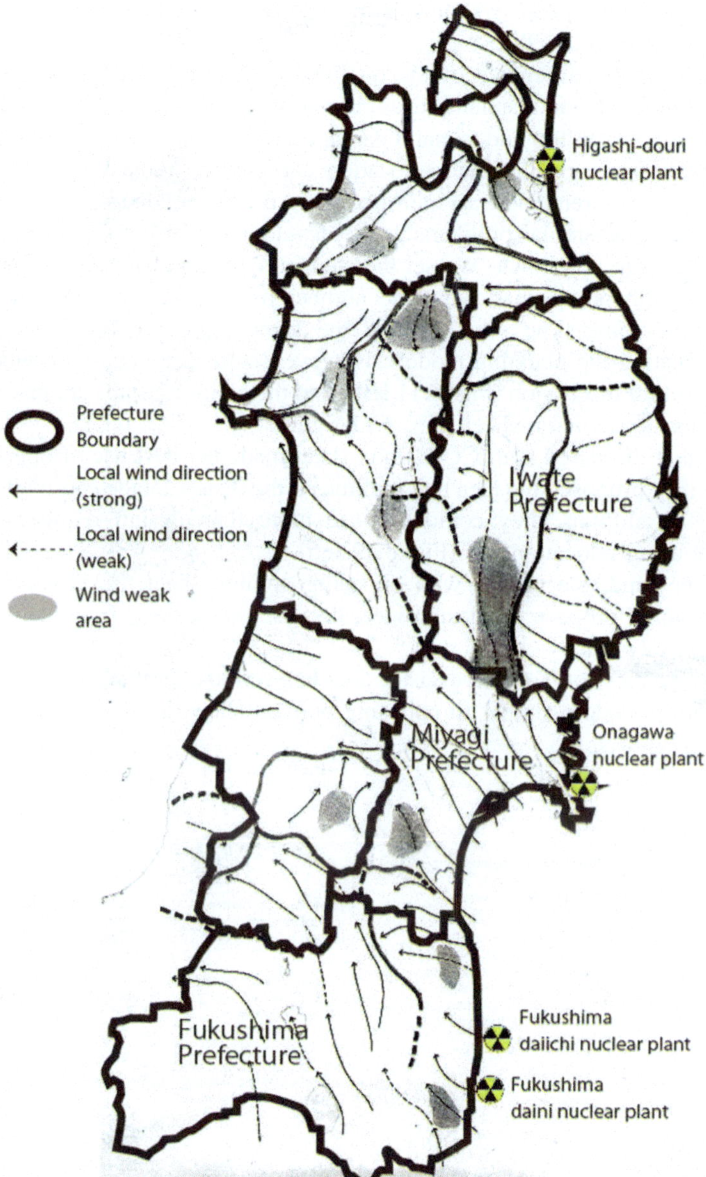

This figure suggests the potential radioactive contamination to Miyagi Prefecture and Iwate Prefecture from the nuclear plants in Fukushima and Miyagi, respectively.

Fig. 3.10 Prevailing wind map of 1980s ecological planning database indicates the potential radioactive contamination diffusion, the author adds prefecture boundary and location of nuclear plants, created based upon data from National Land Agency Japan (1980)

COCD-box (http://www.cocd.org/kennisplatform/cocd-box/) or plasticine, coloured clay material (Roggema 2012b).

We investigated the difference of the quality of reconstruction proposals for 2011 Tohoku earthquake and tsunami disaster area. We offered each working team a new proposal for seawall(s) and tsunami damaged community housing relocation site(s). The proposed seawall(s) and relocation community housing site(s) were required because they remained a controversial issue in the disaster area. This development issue has similar problems as the development of the former nuclear plant and past coastal new town. We set the student's design charrette to work in two situations. Both groups had the same number of students, the same design requirements, the same time schedule and the same plasticine (coloured clay material) for creating the planning models. However, some groups are provided a COCD-box sheet for discussion (Fig. 3.11 left), and the control groups are provided sheets with opposing opinion sheet (Fig. 3.11 right).

The group that discussed with COCD-box sheet made the best use of opposing opinions, and they proposed the idea that minimized the disadvantage caused by the development. In addition, they could propose more than design requirements: seawall and housing relocation site (Fig. 3.12 left).

On the other hand, the groups with opposing opinion sheet (development or preservation) cannot propose the design idea, because they cannot find alternative proposal (Fig. 3.12 right).

Unfortunately, most of Tohoku disaster area has serious conflict, for their discussion process procedure similar to opposing opinion sheet discussion.

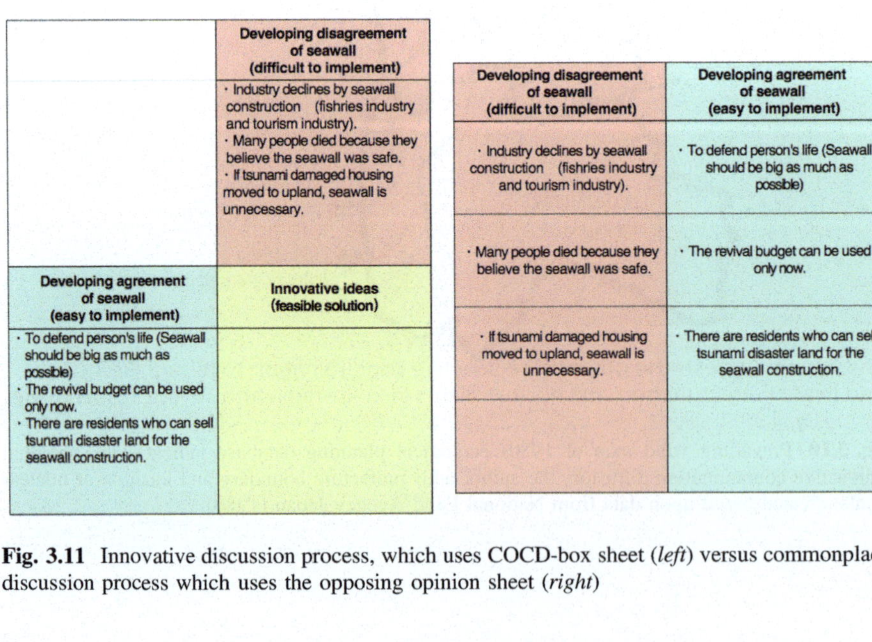

Fig. 3.11 Innovative discussion process, which uses COCD-box sheet (*left*) versus commonplace discussion process which uses the opposing opinion sheet (*right*)

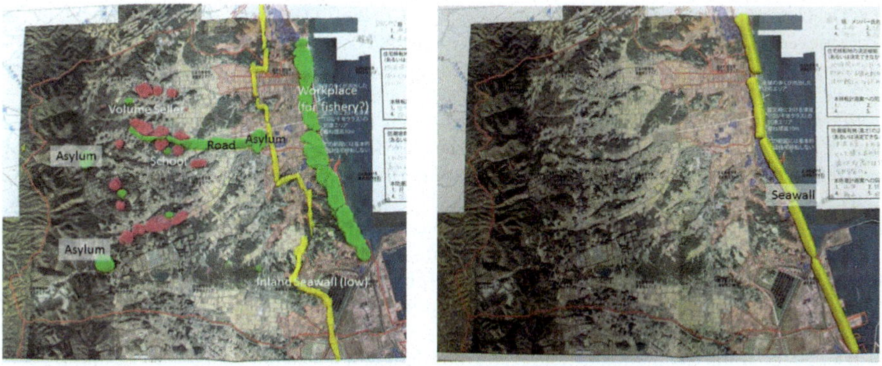

Fig. 3.12 Innovative proposal using COCD-box sheet (*left*) versus commonplace proposal using opposing opinion sheet (*right*)

3.7 Conclusion

As mentioned above, we can learn from an aspect of the landscape, which aims to balance development and preservation, and we can use it for actual planning processes.

This chapter focused on some lessons of man-made failure of the disasters (special planning issue), which can be learnt from this disaster. For the overall consideration to 2011 Tohoku earthquake and tsunami disaster appears insufficient. An East Japan great earthquake seemed actually an unprecedented large-scale natural disaster. However, we think that it was also a human impacted disaster although it hides in the big impact of the natural disaster.

We believe that looking fairly to the common failure of human decision for the disaster brings us an achievement of sustainable development, reduction of social cost, and people's happiness.

Acknowledgements This work was supported by a Grant-in-Aid for Scientific Research (C) Number 24560775, Grant-in-Aid for Young Scientists (B) Number 15K21039. The author gratefully acknowledge the work of past and present members of our laboratory: Makoto Fujii and Gen Shintaku.

References

Agency for Natural Resources and Energy 2011. Dengen Rittchi Seido no Gaiyou-Outline of power supply location institution-, Agency for natural resources and energy. Retrieved from, https://www2.dengen.or.jp/html/leaf/seido/files/richigaiyo-201003.pdf.

All Japan Council of Local Governments with Atomic Power Stations. 1998. Koteishisanzei Syunyuu to Dengenrichi Sokushintaisaku Koufukin-Property tax income and power supply location promotion grant-, All Japan Council of Local Governments with Atomic Power Stations. Retrieved from, http://www.zengenkyo.org/ayumi/koufukin.html.

COCD: Centrum voor de Ontwikkeling van het Creatief Denken. De COCD-box. COCD. Retrieved from, http://www.cocd.org/kennisplatform/cocd-box/.

Dailynewsagency. 2015. 39 pictures with which March of last year and 2013 was compared 2 years ago at the same location of the disaster area of an eastern Japan great earthquake, Retrieved from, http://dailynewsagency.com/2013/03/08/japan-two-years-post-tsunami-3n9/.

Fujii, Y., K. Satake, S. Sakai, M. Shinohara, and T. Kanazawa. 2011. Tsunami source of the 2011 off the Pacific coast of Tohoku Earthquake. *Archives of Earth Planets and Space* 63 (7): 815–820.

Geospatial Information Authority of Japan. 2015a. Landuse transformation archive. Geospatial Information Authority of Japan. Retrieved from, http://mapps.gsi.go.jp/maplibSearch.do#1.

Geospatial Information Authority of Japan. 2015b. 2011 year Tohoku-district Pacific offing earthquake and old and new picture comparison, Retrieved from, http://saigai.gsi.go.jp/h23taiheiyo-ok/hikaku/index.html.

Google. 2015. Mirai heno Kioku, Retrieved from, https://www.miraikioku.com/2015.html.

IRSN: Institut de radioprotection et de sûreté nucléaire. 2011. Simulation de Nuage radioactif Fukushima. IRSN. Retrieved from, https://www.irsn.fr/FR/Actualites_presse/Actualites/Pages/20110319_simulation_dispersion_panache_radioactif.aspx#.WOx9C1OLSRs.

Iuchi, K., L.A. Johnson, and R.B. Olshansky. 2013. Securing Tohoku's future: Planning for rebuilding in the first year following the Tohoku-Oki earthquake and tsunami. *Earthquake Spectra* 29 (Suppl. 1): 479–499.

Maeda, T., T. Furumura, S. Sakai, and M. Shinohara. 2011. Significant tsunami observed at ocean-bottom pressure gauges during the 2011 off the Pacific coast of Tohoku Earthquake. *Archives of Earth Planets and Space* 63 (7): 803–808.

Matanle, P. 2013. Post-disaster recovery in ageing and declining communities: The great east Japan disaster of 11 March 2011. *Geography* 98 (2): 68–76.

McHarg. L.I. 1969. *Design with nature*. New York: The Natural History Press Garden City.

Ministry of Education, Culture, Sports, Science and Technology. 2011. About a result of measurement of an airplane monitoring. Ministry of Education Science and Culture and United States Department of Energy. Ministry of Education, Culture, Sports, Science and Technology. Retrieved from, http://radioactivity.nsr.go.jp/ja/contents/4000/3710/24/1305820_20110506.pdf.

Mori, N., T. Takahashi, T. Yasuda and H. Yanagisawa. 2011. Survey of 2011 Tohuku and Earthquake tsunami inundation and run-up. *Journal of Geophysical Research Letters* 38 (7). DOI 10.1029/2011GL049210.

National Land Agency Japan, Regional Planning Team. 1980. Land use aptitude evaluation technique investigation based on ecological planning, appended figure 5 map with table of degree of risk of disaster (in Japanese).

NAVER. 2015. Disaster area of an eastern Japan great earthquake, Just after the earthquake disaster and the state now, Retrieved from, http://matome.naver.jp/odai/2133066088706857401?page=2.

NHK NEWS Web. 2014. *Shinsai Sannen, Genpatujiko no Songaigaku jyuiicho yen Goeni- To the damage amount of nuclear accident for earthquake seems 11 trillion yen or more, during three years-*, NHK NEWS Web. Retrieved from, http://www3.nhk.or.jp/news/genpatsu-fukushima/20140311/1516_songaigaku.html.

Roggema, R. 2012a. *The design charrette. Ways to envision the future*. Dordrecht, Heidelberg, London, New York: Springer.

Roggema, R. (ed.). 2012b. *Swarming landscape: The art of designing for climate adaptation*. Dordrecht, Heidelberg, London, New York: Springer.

Takeuchi, K., T. Elmqvist, M. Hatakeyama, J. Kauffman, N. Turner, and D. Zhou. 2014. Using sustainability science to analyse social–ecological restoration in NE Japan after the great earthquake and tsunami of 2011. *Sustainability Science* 9: 513–526.

Uehara, M. 2012. Could ecological planning data base on land agency report in 1980 prevent the actual disaster of fukushima nuclear plants caused by the Tohoku earthquake and tsunami? *Landscape Research Japan Online* 5: 28–32.

Uehara, M., T. Inoue, and G. Shintaku. 2015. The favorable settlement relocation process after the 2011 earthquake and tsunami disaster in japan by evaluating site environments and accessibility. *International Review for Spatial Planning and Sustainable Development* 3: 119–130.

Zaré, M.S., and Ghaychi Afrouz. 2012. Crisis management of Tohoku Japan earthquake and tsunami. *Iranian Journal of Public Health* 41 (6): 12–20.

Chapter 4
Government Led Reconstruction Activities in Fukushima with a Specific Focus on the Reconstruction Supporters Project: Importance of Human Recovery

Ryo Sakurai and Wanglin Yan

Abstract After the Great Eastern Japan Earthquake on March 11, 2011, which resulted in a triple disaster (earthquake, tsunami and nuclear accident), the Japanese government both at local and national level has implemented reconstruction and revitalization measures. In this chapter, the author discusses the importance of human recovery after the disaster and specifically examines one project; the Reconstruction Supporters Project. This project initiated the restoration of the psychological health of communities and residents. The Ministry of Internal Affairs and Communications established the project, and has currently arranged 452 supporters to settle in the devastated area helping local residents and to revitalize the region.

4.1 Overview of Efforts by Government and Fukushima Prefecture for Reconstruction

The Great Eastern Japan Earthquake occurred on March 11th, 2011, recorded Magnitude 9.0, the highest on record, and together with the tsunami, devastated the Tohoku region. This disaster led to destruction of the Fukushima Daiichi Nuclear Plant (FDNP), affiliated with the Tokyo Electronic Power Company, with potential hydrogen explosion and resulted in a leak of radioactive substances contaminating the surrounding area in Fukushima Prefecture. On April 22nd, the area inside of 20 km of the FDNP was designated as a “Restricted Area” and the Japanese Nuclear Emergency Response Headquarters (NERHQ) ordered residents living in this area to evacuate (International Atomic Energy Agency 2011). The area in

R. Sakurai (✉)
College of Policy Science, Ritsumeikan University, Osaka, Japan
e-mail: ryo223sak@gmail.com

W. Yan
Faculty of Environment and Information Studies, Keio University, Fujisawa, Kanagawa, Japan
e-mail: yan@sfc.keio.ac.jp

R. Roggema and W. Yan (eds.), *Tsunami and Fukushima Disaster: Design for Reconstruction*, DOI 10.1007/978-3-319-56742-6_4

which an annual dose rate of radioactive substances was more than 20 mSV (micro Sievert) was designated as the "Deliberate Evacuation Area", and the area between 20 and 30 km from the nuclear plant was declared the "Evacuation-Prepared Area" (International Atomic Energy Agency 2011). This led to about 99,000 local residents evacuating from those areas (Asaoka and Ishiyama 2013).

The Japanese government established "the basic policy for reviving from the Great East Earthquake" and proposed "creative revitalization" as the main theme and strategy (Ministry of Education, Cultures, Sports, Science and Technology 2011). By saying "creative revitalization", the government stressed the revitalization as a good opportunity for economic growth of the country, and the Tohoku region could be redeveloped as an area, which could provide goods and sell these to global and large-scale companies. However, there is criticism that this economic focused revitalization delays revitalization and capacity building of local communities (Asaoka and Ishiyama 2013). In other words, revitalization should also focus on energizing communities and reviving networks among residents resulting in "revitalization of people".

Fukushima Prefecture declared in the "Plan for Revitalization of Fukushima Prefecture" that:

1. A safe, secure and sustainable society free from nuclear power should be established;
2. A revitalization should be developed, which brings together everyone who loves and cares about Fukushima;
3. A homeland should be created on which we once again can all be proud of, and regain trust in basic concepts (Fukushima Prefectural Government 2011).

Besides priority projects as environmental restoration, revival of primary industries, and promotion of renewable energy, the prefecture raised "reviving and developing community bonds" including "improving the living and social environment for evacuees", "revitalization of communities together with evacuees", and "building resilient communities through a shared vision of the future" as specific actions and major projects (Fukushima Prefectural Government 2011).

Globally, more efforts for reconstruction by governments have focused on restoration of physical infrastructure such as buildings rather than social aspects such as encouraging people and energizing the communities after a disaster (Chandra and Acosta 2010). This can also be claimed for the Fukushima-case in which a lot of efforts were made for the reconstruction of infrastructure as well as decontamination of radiation (Zhang et al. 2014). A previous study found that psychological impacts and trauma of local residents could last for a long time after the disaster (Adams et al. 2011). Especially, people in Fukushima are likely to have much higher levels of stress than those in Iwate and Miyagi because of their fear towards radiation and the loss of neighbourhoods networks, caused by the evacuation (Zhang et al. 2014).

In the next chapter, the "Reconstruction Supporters Project", which was established to actively enhance capacity and community networks and rebuild confidence and pride of residents, is discussed.

4.2 The Reconstruction Supporters Project by Ministry of Internal Affairs and Communication

In order to revive the community from disaster, not only physical support such as financial aid or developing infrastructure, including buildings is necessary, but also mental support for residents and encouraging people is important (Chandra and Acosta 2010; Itonaga 2012). This is because local people should take the initiative to effectively revive the community and rebuild the society sustainably (Funabasi 2012). Previous studies revealed that community building is not effective or sustainable when it is led by the government and/or outsiders, such as researchers, in a top-down approach (Zanetell and Knuth 2004; Hill 2008). In addition, cases from all over the world show that it is important for the community to have a strong network among residents, the capacity to cope with various local issues and motivated residents, in order to build resilient communities (Tidball and Krasny 2009; Itonaga 2012). However, many local regions in Japan are suffering from aging and depopulation because of reasons such as urbanization, including the lack of jobs in the countryside and the younger generation moving to the cities. Most of the regions where three disasters—earthquake, tsunami, and spread of radioactive contaminants—occurred had those problems also. According to a previous study (Barusch 2011), older adults are the most vulnerable for a disaster and need tangible assistance. The disaster accelerated the aging and depopulation rate and one study revealed that in one city, Minamisoma, in Fukushima, the ratio of elderly aged over 65 increased from 25.92% in March 2011 to 32.88% in March 2014 (Zhang et al. 2014). It can be challenging to revive the community where there are fewer people of a younger generation who can actively get involved in activities. In addition, evacuation has caused a collapse of local communities including daily relationships and networks in neighbourhoods (Itonaga 2012). One study in Iitate Village, a contaminated area in Fukushima prefecture, found that two thirds of the residents who participated in the research either lost or had fewer communications with other villagers (Itonaga 2012). Fujimoto (2013) reviewed efforts of reconstructions after a disaster all over the world and concluded that the personnel necessary for effectively reviving local communities should be able to (1) connect and help local residents as a coordinator and (2) reach consensus among residents for a future plan.

In order to solve these problems and support revitalization of local regions after the disaster, the Ministry of Internal Affairs and Communication established the "Reconstruction Supporters Project" in 2012 (Ministry of Internal Affairs and Communications 2014). Through this project, people who are selected as supporters settle in disaster regions and support local residents by improving their living environments, caring for elderly people, building networks, providing opportunities in which residents can discuss about the plans for the future, and all other activities that enhance revitalization. Those supporters live in the region for a maximum of five years and receive a salary (about 2,000,000/year) and other necessary expenses (Ministry of Internal Affairs and Communications 2014). Currently, 452 supporters are settled in the devastated areas, mainly in three prefectures (Fukushima, Iwate, and Miyagi), support local residents and revitalize the region (Ministry of Internal Affairs and Communications 2014). In the next chapter, two case studies of such a Reconstruction Supporters Team, based at Tamura city located southeast of Fukushima city and at Minamisoma city located east side of the prefecture are presented.

4.3 The Reconstruction Supporter Team at Tamura City and Minamisoma City of Fukushima Prefecture

Tamura city is located southeast of Fukushima city and has a population of 37,658 (11,806 households) (Tamura City 2015). Miyakoji town is located in the east part of the city and some areas of the town are within 20 km from the Fukushima Daiichi Nuclear Plant. Therefore, after the Great Eastern Japan Earthquake, Tamura city asked all residents in this town to leave as some areas were classified as the "Restricted Area". As of June of 2013, 2279 people (666 households) evacuated, which resulted in 950 residents living in temporary housing and 871 residents in rental houses (Iwasaki 2013). However, on 1 April 2013, the government lifted the regulation of Miyakoji town as "Restricted Area" since the contamination rate was low, and declared the area safe for living (Asahi Shimbun Digital 2014; Tamura City 2013). Schools and other facilities started, and some residents started to return and live in the original houses of the town (Tamura City 2013). Since October 2014, 33.1% of the residents (n = 117), whose houses were in the Restricted Area, returned to their original residence (Kohokushinpo Online News 2014).

Although the government declared the Miyakoji area is safe to live, in order to revive the local society as it was before the disaster, resuscitation of the community through reconnection and encouragement of people (who once were evacuated to different locations), is also very important. Providing financial aid and/or re-construct buildings does not necessarily create lively communities unless people living there actually feel happy and energetic (Chandra and Acosta 2010).

Tamura city utilized the "Reconstruction Supporters" program, and established the Tamura City Reconstruction Supporters (written as Supporters ever after) in July 2013, by entrusting the project to the Non Profitable Organization "Coaster" (Tamura City 2013). Currently seven of the supporters live in Tamura city and work to take care of sufferers and to support activities to reconstruct communities. The supporters consist of four members from Tamura city and three from outside of the prefecture, such as Tokyo and Mie, and there are two female members and five males. Most of them are in between 20 and 30 years of age.

Their activities are mainly classified as "supporting residents' lives" and "supporting revitalization of the local community". Activities for supporting residents' lives include removing snow on the roofs of elderly people's houses and clearing the brush around the agricultural fields. Since it snows heavily in Miyakoji Town during the winter period, removing snow on roofs and streets is hard work, and challenging especially for elderly people. Meanwhile, clearing the brush around the agricultural fields and houses is important not only to prevent agricultural damage by wildlife but also to preserve the traditional scenery of the area. Supporters also make leaflets; "Miyakoji is cool! (Iine Miyakoji)" which introduces events and people of the town, and distribute them to all households in the town every month (Tamura City 2013).

As for supporting the revitalization of the local community, supporters provide opportunities in which local residents can think and discuss about what kind of communities they want to build and what they want to do in the town. Supporters also invite volunteers and experts, so that local residents can get new ideas. Enhancing communication and collaboration between local residents and people from outside is one of the goals of the supporters, and since October 2013, supporters and local residents accept volunteers. Volunteer activities are classified as "know about Tamura city", "interact with residents", and "promote attractiveness of the city". Until August 2014, a total of 182 volunteers came to Miyakoji town both from inside and outside the prefecture. (Tamura City 2013).

The Reconstruction Supporters Team is also located in different areas of Fukushima Prefecture including Minamisoma city. Minamisoma city is located in the east coast of the prefecture and has a population of about 63,000 (Minamisoma City 2015). The southern part of the city, Kodaka district is located within 20 km from the nuclear plant and, similar to Miyakoji district, was once listed as the "Restricted Area". Currently, there are three Reconstruction Supporters at Minamisoma city (Fukushima Cooperative Reconstruction Center 2015). Four main activities that the Reconstruction Supporters of Minamisoma implement are:

1. Promoting Minamisoma through introducing about the tradition of the area and selling special products of Minamisoma (such as goods of traditional horse festival; "Somanoumaoi", of the region),
2. Organize and implement guide tours visiting sites damaged by tsunami and explaining incidents and impacts of explosion at nuclear power plant,
3. Supporting to organize and host the traditional horse festival; "Somanoumaoi", and promoting this festival, and
4. Supporting to invent special goods of Minamisoma such as developing pear juice at Kashima district.

Difference between the Reconstruction Supporters activities between Minamisoma and Tamura is that in Minamisoma, activities are more likely focused on tourism while in Tamura, activities are focused on directly supporting individual local residents (such as snow shoveling, visiting each household). While promoting tourism has certain effects on reviving the local economy and encouraging people in the region, as more residents come back to districts from evacuation, activities that focus more on supporting individual residents' lives might be needed. However, activities that directly support residents require time and manpower, and having only three Supporters, it might be difficult for the Reconstruction Supporters Team in Minamisoma to expand and increase their activities. In addition, while the

Reconstruction Supporters Team in Tamura belongs to the city government, the team of Minamisoma belongs to the Tourism Association and these can be one of the reasons why the activities are different. Needs and demands from residents and local stakeholders would change as the reconstruction proceeds and as time passes after the disaster, and for the Reconstruction Supporters Team, understanding these needs and reflecting them would be important for implementing effective supporting activities (Inagaki 2015; Sakurai et al. 2016).

While such activities as re-construction of buildings and decontamination can result in visible and/or measurable outcomes, the impacts that supporters' activities generated to the local communities are hard to see. However, each of these activities are critical first steps for improving residents' psychological well-being through decreasing the mental stress that comes from isolation in the network. Eventually this would help revitalize the people and communities.

4.4 Conclusion

After the disaster, government and local municipalities have done various efforts for reconstruction of the devastated areas. Many of these activities are focused on physical reconstruction such as building infrastructures, however, in order to achieve effective and sustainable revitalization, supporting local people and building capacity of communities are indispensable. The Reconstruction Supporters' efforts and activities provide an example how local people could be encouraged and local communities can be revitalized by the people themselves and by people's interactions.

A previous research revealed that, in the area of Tohoku region where the reconstruction has been on the process, there are not enough experts who understand needs of local residents while there are experts that that reflect government's needs (Suzuki 2014). The Reconstruction Supporters have vital role in bridging local residents and government making sure that residents' needs and opinions are heard and reflected in the policy. Meanwhile, in order to effectively implement activities that meet the demands of residents and foster the reconstruction of the community, the Reconstruction Supporters would need to continuously monitor the condition of the local community (e.g., needs of local residents, residents' evaluation of activities and reconstruction process) (Sakurai et al. 2016). Public survey conducted in Miyakoji town of Tamura city revealed that local residents wanted the Supporters to help residents organize the festival and public meeting (Sakurai et al. 2016). Conversely, activities such as the Reconstruction Supporters visiting each household door to door asking their needs are not so much demanded by residents. The survey revealed that four years after the disaster, local residents' needs are changing; they preferred indirect support (e.g., helping to organize the public meeting in which residents take initiative to implement, helping residents to play a major role in the festival) rather than direct support (e.g., helping each residents by visiting houses, snow shoveling) (Sakurai et al. 2016). Similar situation could likely happen in other reconstruction sites including Minamisoma city. Residents' demands and needs will likely change as reconstruction progresses, and by understanding these demands constantly, and reflecting it to actual supporting activities, the Reconstruction Supporter's efforts could be more effectively help the whole revitalization of the community.

Acknowledgements The author would like to thank the Fukushima Cooperative Reconstruction Centre (Fukushima Renkei Fukko Center) for providing valuable information regarding the Tamura City Reconstruction Supporters and supporting research in Tamura City. The author also would like to thank the Sanaburi Foundation as one part of this study (Sakurai et al. 2016) won the Award for the Most Useful Paper toward Reconstruction from Disaster and received funding from this foundation in 2016.

References

Adams, V., S.R. Kaufman, T.V. Hattum, and S. Moody. 2011. Aging disaster: Mortality, vulnerability, and long-term recovery among Katrina survivors. *Medical Anthroplolgy* 30 (3): 247–270.

Asahi Shimbun Digital. 2014 First lift of restricted area at Tamura city of Fukushima in April, three years after the disaster. (In Japanese). Retrieved February 19, 2015, from http://www.asahi.com/articles/ASG2R5HNKG2RUTIL00X.html.

Asaoka, Y. and T. Ishiyama. 2013. Introduction: Environmental education point of view after the Great East Japan Earthquake. In *Environmental education after the Great East Japan Earthquake*, ed. The Japanese Society of Environmental Education, 1–14. Tokyo: Toyokan Press.

Barusch, A.S. 2011. Disaster, vulnerability, and older adults: toward a social work response. *Journal of Gerontological Social Work* 54: 347–350.

Chandra, A., and J.D. Acosta. 2010. Disaster recovery also involves human recovery. *Journal of the American Medical Association* 14: 1608–1609.

Fujimoto, T. 2013. Perspective on community policy for disaster-stricken region facing population decrease—Based on the result of "revitalization support coordinator" allocation pilot program. *Sustainability Research* 3: 135–149. (In Japanese).

Fukushima Cooperative Reconstruction Centre. 2015 Fukushima Fukkou Ouentai: Minamisoma City. (In Japanese). Retrieved January 3, 2016, from http://fukushimafukkououentai.jp/minamisouma.

Fukushima Prefectural Government. 2011. Plan for revitalization in Fukushima Prefecture (First Version): Forming a link to the future, beautiful Fukushima. Retrieved January 11, 2015, from http://www.pref.fukushima.lg.jp.e.od.hp.transer.com/sec/11015b/fukkoukeikaku1131.html.

Funabasi, H. 2012. Establishing structure for reconstruction. *Trend of Science* 10: 88–90. (in Japanese).

Hill, C.M. 2008. Working with communities to achieve conservation goals. In *Wildlife and society: The science of human dimensions*, ed. M.J. Manfredo, J.J. Vaske, P.J. Brown, D.J. Decker, and E.A. Duke, 117–128. Washington, DC: Island Press.

Inagaki, F. 2015. Yell from Chuetsu to Tohoku. *Sekai* 4: 101–109. (in Japanese).

International Atomic Energy Agency. 2011. Report of Japanese Government to IAEA Ministerial Conference on Nuclear Safety—Accident at TEPCO's Fukushima Nuclear Power Stations. Retrieved February 13, 2015, from https://www.iaea.org/newscenter/focus/fukushima/japan-report.

Itonaga, K. 2012. The great earthquake disaster, the Nuclear plant accident, and regeneration of the disaster area: A report from Iitate, Fukushima. *Kyosei Studies* 5 (1): 1–24. (In Japanese).

Iwasaki, D. 2013. Introduction of Tamura City reconstruction supporter project. Non Profitable Organization Coaster. (In Japanese).

Kohokushinpo Online News. 2014. Not many people returned at east Miyako in Tamura half a year after the lift of evacuation order. (In Japanese). Retrieved February 19, 2015, from http://www.kahoku.co.jp/tohokunews/201410/20141002_61011.html.

Minamisoma City. 2015. Minamisoma City. (In Japanese). Retrieved January 3, 2016, from http://www.city.minamisoma.lg.jp/.

Ministry of Education, Cultures, Sports, Science and Technology. 2011. Image of creative reconstruction. (In Japanese). Retrieved February 18, 2015, from http://www.mext.go.jp/b_menu/shingi/chukyo/chukyo9/shiryo/__icsFiles/afieldfile/2011/06/20/1307068_6_1.pdf.

Ministry of Internal Affairs and Communications. 2014. Reconstruction supporters. (In Japanese). Retrieved December 29, 2015, from http://www.soumu.go.jp/main_sosiki/jichi_gyousei/c-gyousei/02gyosei08_03000067.html.

Sakurai, R., K. Okuda, and D. Tsukahara. 2016. Local residents' perceptions of the Reconstruction Supporters and current situation of reconstruction: Survey toward residents at Miyakoji area of a Tamura city, Fukushima prefecture. *Journal of Rural Planning* 35 (3): 389–397. (In Japanese with English abstract).

Suzuki, T. 2014. Regional community revival through town recovery promoters. *JABS* 127 (1629): 2–3.

Tamura City. 2013. Tamura governmental bulletin. No. 115. Fukushima: Tamura City. (In Japanese).

Tamura City. 2015. Tamura city of Fukushima prefecture. (In Japanese). Retrieved February 19, 2015 from http://www.city.tamura.lg.jp/.

Tidball, K.G., and M.E. Krasny. 2009. Chapter 7: From risk to resilience: What role for community greening and civic ecology in cities? In *Social learning: Towards a sustainable world*, ed. A.E.J. Wals, 292–299. Netherlands: Wageningen Academic Publishers.

Zanetell, B.A., and B.A. Knuth. 2004. Participation rhetoric or community-based management reality? Influence on willingness to participate in a Venezuelan freshwater fishery. *World Development* 32: 793–807.

Zhang, H., W. Yan, A. Oba, and W. Zhang. 2014. Radiation-driven migration: The case of Minamisoma City, Fukushima, Japan, after the Fukushima nuclear accident. *International Journal of Environmental Research and Public Health* 11: 9286–9305.

Chapter 5
The Design Process

Rob Roggema and Wanglin Yan

5.1 Introduction

The municipality of Minamisoma was hit by a double disaster. The tsunami caused a heavy flood in the area and the break down of the nuclear plant of Fukushima caused a severe radioactive pollution. Both impacted the population of Minamisoma. Part of the residents died in the disaster, many left the area if they could while other were forced to move because of the dangerous levels of radiation. Even years after the disaster many people haven't returned to their homes. On the one hand side because they have found a base elsewhere or it is too dangerous still, but on the other hand people are not attracted anymore and have no good reason to return. Apart from government planning for a robust coastal protection wall, there aren't too many initiatives to turn the negative feeling around. Therefore the initiative was taken to organise a design week in Minamisoma in order to give local residents the opportunity to participate in thinking about their future and get the positive feeling a meaningful future lies ahead. The design week took place in September 2014 and was organised in collaboration between KEIO University in Japan and Swinburne University of Technology in Australia.

A similar design week was organized in Kesennuma in 2015. This coastal town suffered from severe flooding as result of the tsunami (Fig. 5.1).

R. Roggema (✉)
Faculty of Design, Architecture and Building, University of Technology Sydney, Ultimo, NSW, Australia
e-mail: rob.roggema@uts.edu.au

W. Yan
Faculty of Environment and Information Studies, Keio University, Fujisawa, Kanagawa, Japan
e-mail: yan@sfc.keio.ac.jp

R. Roggema and W. Yan (eds.), *Tsunami and Fukushima Disaster: Design for Reconstruction*, DOI 10.1007/978-3-319-56742-6_5

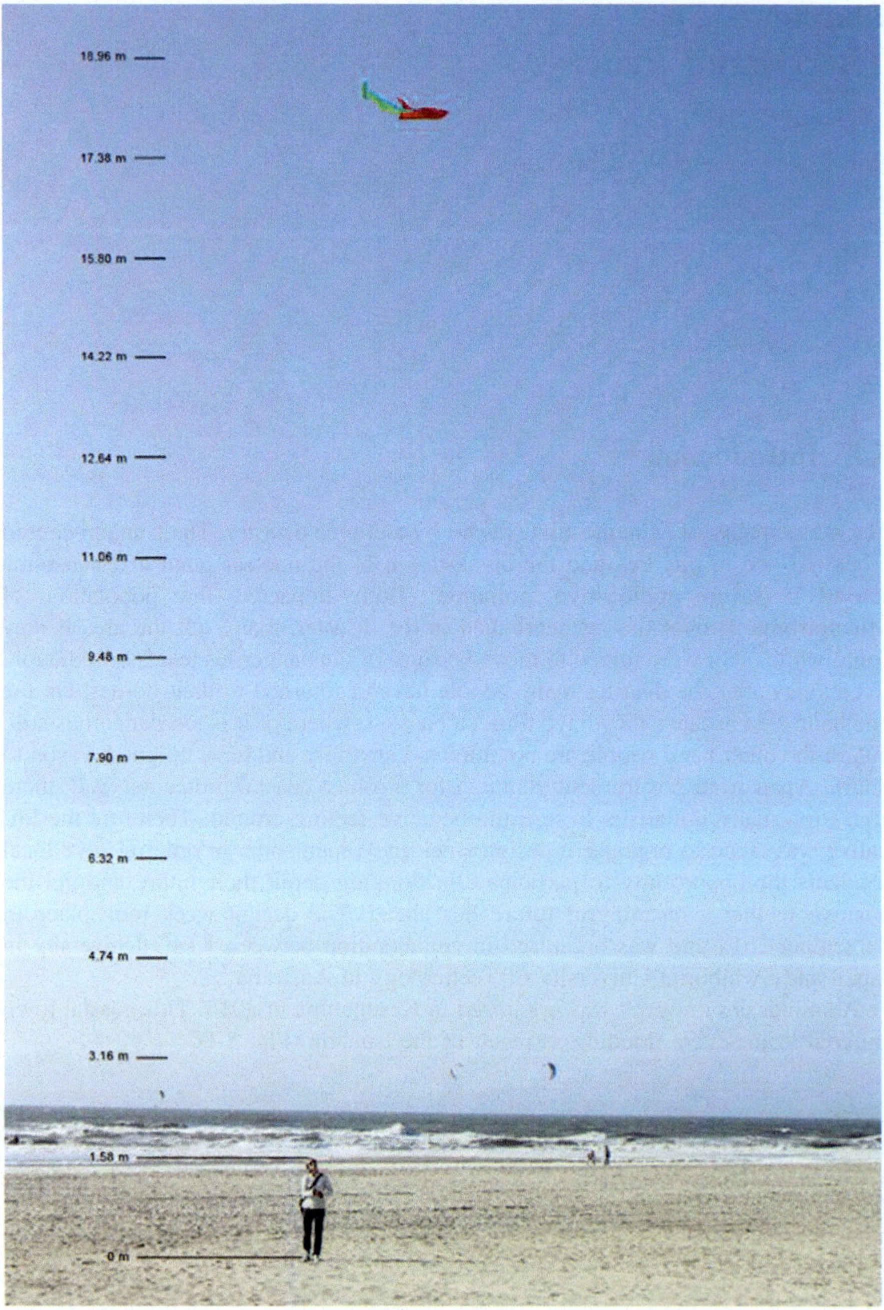

Fig. 5.1 The height of the tsunami on the beach of Minamisoma

5.2 The Design Charrette Process

As mentioned elsewhere, the situation after the disaster consisted of a complex of problems, which interrelated and were difficult to solve in a traditional way. The automated response of rebuilding houses and increase the technical level of protection and solution did not work as the solution of one problem could cause new problems in another. For instance rebuilding houses in a certain area could expose the inhabitants to dangerous levels of radiation. This complexity can only be solved by an integrated approach, directed from a long-term future thinking perspective. The spatial overlap of problems and potential solutions implied also these integrated solutions had to lead to allocated spatial uses. Moreover, future developments and choices could count on the interest of multiple disciplines and a diversity of stakeholders. These were the reasons why a design charrette was a very applicable method to use in this process.

Charrettes have been used successfully in controversial and complicated design and planning problems (Roggema 2013). Examples of these, as given by the NCI (National Charrette Institute) include (Lennertz and Lutzenhiser 2006):

- High stakes projects involving substantial public and private investment;
- Volatile yet workable political environments—situations that are “hot” but manageable;
- Complex design problems;
- Real projects that include imminent development.

The NCI defines the charrette as: “a collaborative design and planning workshop that occurs over four to seven consecutive days, is held on-site and includes all affected stakeholders at critical decision-making points” (Lennertz and Lutzenhiser 2006). Building on this Condon formulates it as: “a time-limited, multiparty design event organized to generate a collaborative produced plan for a sustainable community” (Condon 2008).

A typical Charette consists of several phases Condon (2008), Lennertz and Lutzenhiser (2006):

1. The preparation phase. In this phase everything that is required to hold a successful design charrette is considered. Besides the obvious, required material, bookings of the venue, the main issues in the preparation phase are the design brief and the selection of participants. In the design brief the assignment is clearly defined. The goals and objectives, the design principles, if possible quantitative information, and the performance targets are described. The design brief is written in collaboration with several of the local stakeholders. The selection of participants is critical. People need to be interested in the charrette process. As it comes to the selection of participants, the right “mix” of people is essential. A combination of scientists, local experts and stakeholders, decision makers and knowledge brokers, designers and technical experts, all contribute to the dynamic of the event.

2. The visioning charrette. During this charrette the main goal is to envision the desired future. A typical visioning charrette includes the following parts: an opening event, a site tour, the design stages, during which iterative phases of conceptualisation, drawing alternatives and refinement of the vision take place (or in the words of Condon: talk-doodle-draw), the public meetings and finally the afterparty. In general, this type of charrette involves (mainly) designers for a full week. The purpose is to shift from nice sounding prospects to real solutions, which, in this phase, are represented in designs. During the charrette a common language for solutions is developed and, because no implementation questions will be raised at this stage, the risk ideas will be rejected is minimal. However, it may be expected that, in the openness of the process policy contradictions will be revealed.
3. The implementation charrette. This charrette typically lasts for four days and involves (mainly) design facilitators and stakeholders. Its aim is to develop a shared understanding of the desired future and what is needed to realize this future. In many governmental organizations the so-called "window-of-no" is in operation. This prevents change from happening and is often well established through unwritten codes and invisible agreements. The implementation charrette is a powerful tool to go past this window of no. The fast and efficient charrette process involves stakeholders in a powerful integrative way. Because the participants are connected and the solutions cannot be seen as separate, the charrette process helps to outpace approval processes, which sometimes may take years.
4. The writing of the charrette report. In the report the results of the charrete are integrally presented in a visual and clear way. The report functions as the "contract" for the participants and may be used in formal decision making processes.

In conducting design charrettes Condon (2008) defines nine general rules for a good process. The four acknowledged as the most significant are:

a. Design with everyone: Despite the fact that becoming a designer requires thorough training and very specific skills, the design process as undertaken during charrettes is integrative and contains a variety of possible solutions. This is partly an intuitive and judging activity, which makes it accessible for many individuals. In this sense, everyone is a designer;
b. Start with a blank sheet: If the group of participants are standing around the table, on which a large map of the site is laid down, the simple action to overlay this map with a blank piece of transparent paper will do. The invitation and the challenge lie before all. Everyone is invited to fill in the future and a shared vision will, in the hours to follow, fill up the formerly empty paper;
c. Provide just enough information: Too much information causes decision paralysis and too little produces bad proposals. Just enough is mainly arranged through the expertise of the participants and will be provided during the charrette in a concise and accessible manner (maps, schemes);

d. The drawing is a contract: All drawings produced during the charrette embody the consensus as experienced and achieved by the charrette team. They form a well-understood agreement, or contract, in images amongst the group. The drawings cannot be broken without consent of the group and function as such as a very strong commitment.

The Minamisoma and Kesennuma design charrettes were set up as a visioning charrette.

Before this visioning charrette took place, the design brief was developed and send to the participants. In the design brief the design task, the problem and background information was brought together. The design brief comprised also of a list of participants and a detailed program.

The length of the program was six days, consisted of several stages, contained intensive contact with local leaders and residents of the areas, and students of the local agricultural high school at certain moments.

The process of the charrettes started with several site visits. The core team of designers, which consisted of experts and academics from the Netherlands, Australia and Japan and a group of post-doc and doctoral students of KEIO University, recruited from different countries, visited several locations in the area. After the site visits the 30–30 workshop was held with residents and high school students. This workshop explored the knowledge of crucial happenings and knowledge of the participants about the past (30 year ago) and the future (30 years ahead), for four themes: population, environment, economy and culture. The results of this workshop formed the input for the first stage of designing in design teams. The designs that were created responded to the task of creating resilience in the landscape recovering from the nuclear disaster and/or tsunami. Moreover, the designs encapsulated thinking about the future. How could this landscape become resilient for future disasters, or in other words: could the landscape anticipate future change in its landscape design in order to navigate around eventual disasters, or turning these into benefits. After three days of design work the design team presented their results to the group of residents and high-school students. The presentations formed the introduction of a workshop in which the participants themselves worked on future visions for the area. They build 3D-models with plasticine to show their desired future. At the end of this workshop the groups presented their work and conclusions were drawn.

5.3 Facilitation

Besides the content of the design charrettes the facilitation and organisation is a crucial factor to make a charrette a success. The timing of events, the venue and the way the design team could relate to local people, habits and get a feel for the regional surroundings are all very important. The design charrettes in Kesennuma and Minamisoma each consisted of the following aspects:

1. Duration: the length of the charrettes was 7 days, which is long enough for participants who were not familiar with the area to experience and see the area, and to understand the problems and opportunities. It gave also space to a substantial design process with participation of local leaders, juniors and other people.
2. Extensive site visits: the design team has, in each design charrette, visited the area extensively. The surroundings were shown, local entrepreneurs, social groups and other stakeholders were visited and who, in several occasion, presented their interests, history or experience of the area. Specific sites were explored, where the tsunami has devastated a particular location, or where certain problems for rebuilding occur (Fig. 5.2).
3. Local experiences: The design team was offered to indulge the traditional culture. As a means of building an emotional basis for the design work and to better understand the local context the team was hosted in a traditional hotel and had dinner with a traditional Japanese barbecue (Fig. 5.3).
4. Contact of facilitators with participants: The members of the design team came from across countries, disciplines and study levels, so it was important to deepen the contact during the first days of the charrette. The team leaders and

Fig. 5.2 Site visit in the paddy fields of Minamisoma (*photo* Rob Roggema)

Fig. 5.3 Traditional Japanese Barbeque (*photo* Rob Roggema)

experienced team members took care of the younger, less experienced members and helped them to understand the problem, the program and expectations. Besides informal conversations helped everyone to get comfortable with the approach of the design charrette.

5. Contact of design team with local people: During the design charrettes several contact moments with local groups, such as high school students, local leaders and social groups were included in the program. Through an intensive way of group-oriented work, the design team and the participants learned from each other and became collaborators, which made a joint co-creative process possible.
6. Uniqueness of the venue: the venues of the design team were unique. In Minamisoma the space to use was a part of a local supermarket (Fig. 5.4) and in Kesennuma a space was used in an elderly and disabled people's home. This made the work feel less like 'another day at the office', because the occasional contacts with people visiting the supermarket of living in the elderly home, caused an awareness and linkage with the local situation.

Fig. 5.4 Workshop space in the local supermarket, Minamisoma (*photo* Rob Roggema)

5.4 Step One: Understanding

The first stage in the design charrettes is to gain understanding of the local situation, the problem, the power balances and the spatial opportunities. As mentioned above the site visits were a crucial way of obtaining information from the ground level. Not only took the design team a look at the area, presentations of local stakeholders were given and numerous conversations with local people occurred.

The second part of gaining understanding of the area consisted of the so-called 30–30 exercise (Fig. 5.5). In this exercise the participants needed to identify the most important issues and problems of 30 years ago on four themes: population, economy, culture and environment (see Table 5.1). To distinguish the topics the people were divided in four groups and were given 7 min to write down as many topics for one theme. After seven minutes the groups rotated to the next theme and added to the list produced by the earlier group. This was repeated until every group added topics to every theme. At the end the last group sitting at each theme needed to select a top 5 of issues that were presented to the entire group. After this 30 years back exercise was finishes, exactly the same happened, but then looking ahead in the future for 30 years. In Minamisoma the group consisted of local leaders and a class of high school children, while in Kesennuma a social group of elderly people participated in this exercise. The results (Table 5.1) showed a broad interest in the themes concerning the future and a thorough understanding of the issues of the area. The results also make clear that it is for all participants well possible to think for out in the future. The set-up of this exercise helped the people to do this because they

Fig. 5.5 Working on the 30–30 exercise (*photo* Rob Roggema)

realised that the world of 30 years ago was a complete different one than the one they know now. This implied for them the huge transformations that will undoubtedly occur in the upcoming 30 years, and this made it easier for the participants to imagine the future.

The results of this 30–30 exercise as well as the insights gained during the field trips were used in the design process. The outcomes form the basis for the designs, as they inform the design team not only about the encountered problems in the area, but also suggest directions for the future.

5.5 Step Two: Creation

In the second stage of the design charrette the creation of design propositions is central. In both charrettes, in Minamisoma and Kesennuma, the design team was dived in two sub-teams, who each developed their own design for the area. The design team consisted of experienced researchers and designers from Japan, Australia and the Netherlands, and students with different backgrounds from China, Japan, Iran. This provided a mix of experiences, disciplines, backgrounds, age and interests hence multidimensional teams could be created.

The design process itself took place in two stages. The first stage is the interpreting, drawing and sketching stage. The information, which was collected in the first stage, and additional literature and expertise of the design team is used to

Table 5.1 Overview over the results of the 30–30 exercise in Minamisoma

	Population	Economy	Culture	Environment
30 years ago	Bigger population than today	Economy was growing	Overseas travel became popular	Increase in pollen allergies
	Population was balanced	Daily life was vibrant	Aum Shinrikyo cult group and subway sarin incidents	Serious pollution of urban rivers
	Large families were normal	High quality of industrial products appeared	Nuclear power was operating	Weak regulation of agricultural chemicals
	Many were farmers	Entrepreneurial spirit	Cars, telephones	No air conditioners at home
	Many were living/working away from home	Transportation was convenient	Super Nintendo Entertainment System	Rampant resort development
	Communities active in the region	Greed was rampant	Video tape	Standard of living was rising
30 years ahead	Automated hospitals	Self-sufficiency put into practice	Noma-Oi traditional horse festival continues	Fukushima Daiichi nuclear accident has been dealt with
	Compact cities are possible	Almost 100% of homes have solar power	Famous worldwide as a horse town	CO_2 emissions gradually declining
	Elderly are able to care for each other	People produce electricity by themselves	City returns to conditions before the disaster	Effects of radioactive materials have disappeared
	Super-aged society, average age 95	Gasoline is 1000 yen per liter	Regular international events on theme of horse races	Global warming continues
	More farmers due to agricultural renovation	Other construction materials have replaced concrete	Noma-Oi traditional horse festival recognized as World Heritage	Renewable energy is a part of everyday life
	World-class health facilities/operations for elderly	Possible to live on farming	More new events	An increase in various kinds of disasters

design an area which is sustainable, resilient and adaptive, not only in the present, but especially in the farther future. The cultural history of the location, in combination with adaptive design principles, such as the flexibility of the coastal zone, the provision of renewable energy and local food production, lead to regional designs.

In the case of Minamisoma the link with the yearly horse festival, the staged extensions of the area in which food can be produced after it is declared safe, the dissipative coastal structures defending the coastline for future tsunami's using algae fields all together form a plan (Fig. 5.6) which relates to the local context and is ambitious enough to deal with future uncertainties.

The same methodology is used to design a plan for Kesennuma (Fig. 5.7). Here, the cultural heritage of local stories and reigns is used to design a touristic system for the area behind the tsunami boundary. This is the line in the landscape where the tsunami reached inland. In the area between this line and the ocean a dissipative zone is created where water and land intermingle in order to minimise the power of eventual future flooding.

The second phase of the design process consisted of plasticine building. The design team created a 3D model on the basis of the drawings. Working with plasticine is beneficial for the collaboration in the design teams. When working and building with their hands the participants lose their rationality and leave a protective yet defensive attitude. Instead they discuss the built elements team members propose and add or amend the plasticine instantly. The models also show a 3D image of the future as the design team sees it. The built models of Minamisoma (Fig. 5.8)

Fig. 5.6 Design for the Minamisoma area (Rob Roggema, Luke Middleton, Wanglin Yan and Akihiro Oba)

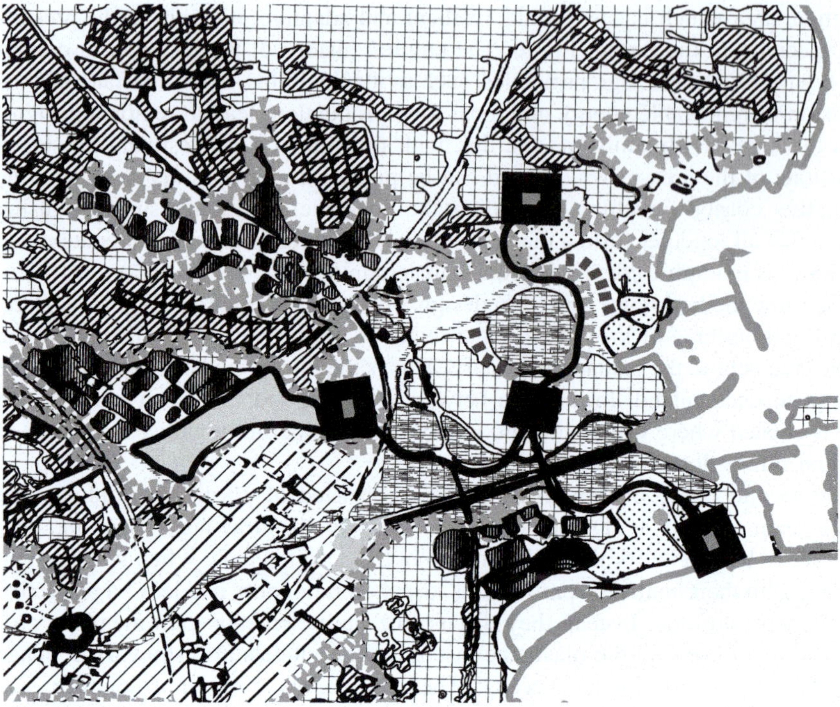

Fig. 5.7 Design for the heart of the Kesennuma area (Rob Roggema, Wanglin Yan, Luke Middleton)

Fig. 5.8 Plasticine model of the Minamisoma plan (*photo* Rob Roggema)

Fig. 5.9 Plasticine model of the Kesennuma plan (*photo* Rob Roggema)

and Kesennuma (Fig. 5.9) give herewith a colourful impression of the future potential of the area.

5.6 Step Three: Co-creation

The third stage of the design charrette is the co-creation phase. In this phase local people, inhabitants of the area, stakeholders and local leaders design their own future, based on the propositions of the design teams.

This co-creation already started in the 30–30 workshop in the beginning of the week, where high school students and local leaders (in Minamisoma) and the social group of elderly (in Kesennuma) identified the major historic trends and the ambitions for the future.

The main part in this stage is however the co-design, which took place at the end of the design charrette. In Minamisoma and Kesennuma the meeting started with brief presentations of the results of the design teams. The analyses, concepts, drawings and plasticine models were shown as an example and inspiration to the audience. After this the participants were divided in mixed groups and were asked to build their own spatial future for their own area in a plasticine model. During the process of building the model, strong collaboration took place and people from different backgrounds and ages jointly and coherently came to one spatial proposition. Because of the composition of the groups, and the fact plasticine requires everyone to become an active builder, the process of building together takes shape very easily. All participants are involved and even fanatic about finishing their models. This attachment to the product implies also that people will hold on to their outcomes and this gives a great way of shared support for future planning.

This process of co-creation was finalised with presentations of the groups' work in front of a panel with decision-makers. This moment forced the groups to finalise their proposals on time and come up with a serious design propositions, which they all did.

5.7 Conclusion

In an area that has to recover from a serious disaster, such as the 3–11 tsunami in Japan, generally a process of rebuilding will start. This rebuilding encompasses the physical removal of debris and other impacts of the disaster, implementation of temporary housing and amenities and the planning for a new future. This planning often resembles the past, and many 'new' buildings and environments seem to be copies of what was before. In most cases, and also in the Japan reconstruction, the government takes the most important position as the leading entity of the planning process. Often, citizens are left out of the planning process and only informed about the plans. Especially in such complex contexts as the reconstruction area in Japan, but also suitable for planning processes deemed 'regular', the design charrette could be a suitable way to involve society and give citizens a well-received voice in the planning of the future of their direct environment.

When working in a design charrette context it is beneficial for the end-result if the whole process is cut in three stages. Every stage has its own typical methods and challenges. In the first stage of 'understanding' is characterised by access to formal and informal information about the area, the problems and the opportunities. During the second stage of 'design' the collected information and expertise is used to in-depth designing of the site. The design team usually consists of a mix of expertise, background and age, and proposes design propositions that are integrated, comprehensive and imaginative. The use of tactile methods, such as the building of plasticine models is essential in creating unexpected views on future development. The third stage of 'co-design' is the phase where citizens, local key players and others build on the design propositions to redesign the area according their own wishes and demands. This phase modifies the earlier propositions and by using similar tactile methods the people involved the support for the final design outcomes is guaranteed.

The approach of design charrettes implies an intensive way of working together. The design team is working together for 3–4 days in a row to come up with relevant yet inspirational designs. This requires a positive attitude of everyone, a challenging task and a beneficial result in terms of satisfaction, experience or adventure. In the stages when citizens and key players are involved the design charrette is at its best because true co-creation emerges when people are collaborating in an intensive, creative and demanding way. The results of the charrette resemble co-creation. There is no sole designer, it is team-work that defines the success of the outcomes.

References

Condon, P.M. 2008. *Design charrettes for sustainable communities*. Washington, Covelo, London: Island Press.

Lennertz, B., and A. Lutzenhiser. 2006. *The charrette handbook. The essential guide for accelerated collaborative community planning*. Chicago: The American Planning Association.

Roggema, R. (ed.). 2013. *The design charrette: Ways to envision sustainable futures*, 335. Dordrecht, Heidelberg, London: Springer.

Chapter 6
Planning and Design in Minamisoma: Reborn, Rethink, Return

Rob Roggema, Luke Middleton and Wanglin Yan

6.1 Background

The Great Eastern Japan Earthquake in March 2011 and subsequent Tsunami (Fig. 6.1) resulted in a substantial area that was flooded (Fig. 6.2) and a nuclear disaster in the Fukushima Daiichi Nuclear Power Plant (FDNPP) affiliated to Tokyo Electric Power Company (TEPCO) (Suzuki and Kaneko 2012). This accident was recorded as one of the most devastating nuclear disasters in human history. In the aftermath of the earthquake, the tsunami inundated the power plant, causing power failure and the breakdown of the reactor cooling system. The overheating reactors started to melt down and caused a hydrogen-air explosion, releasing a substantial amount of radioactive substances, including 1.6Å–1017 Bq for 131I, 1.5Å–1016 Bq for 137Cs. This caused a contamination of an area of more than 1700 km^2. 146,520 people from Fukushima Prefecture had to be evacuated. The nuclear explosion was rated Level 7 in the International Nuclear and Radiological Event Scale (INES) of the International Atomic Energy Agency (IAEA), the most severe one since the Chernobyl accident.

The Japanese central government initiated ambitious reconstruction strategy soon after the disaster, revitalizing the region through *creative reconstruction*.

R. Roggema (✉)
Faculty of Design, Architecture and Building, University of Technology Sydney, Ultimo, NSW, Australia
e-mail: rob.roggema@uts.edu.au

L. Middleton
EME, Melbourne, Australia
e-mail: luke@emedesign.com

W. Yan
Faculty of Environment and Information Studies, Keio University, Fujisawa, Kanagawa, Japan
e-mail: yan@sfc.keio.ac.jp

R. Roggema and W. Yan (eds.), *Tsunami and Fukushima Disaster: Design for Reconstruction*, DOI 10.1007/978-3-319-56742-6_6

Fig. 6.1 Image of the tsunami hitting Minamisoma

Correspondently, local governments adopted reconstruction plans for a sustainable and resilient society. Though five years have passed since then the affected area is still undergoing recovery. There is a big gap between hope and reality, especially in the radiation-contaminated areas where the reconstruction efforts were not progressing as quickly as those in areas that were only affected by the earthquake and tsunami. People were seriously concerned about the risk of radiation and the future of the region (Zhang et al. 2014).

A participative design process can be suitable in such circumstances, where the public is anxious and the solutions seem to be delayed and complex (Luz 2000). A design process provides opportunities for participants to share feelings and look for common visions. A participative process stimulates and enhances the ability of local communities to absorb sudden change, to cope with devastation and damage, and to organize their own resources, hence increasing resilience.

Meanwhile, several requirements have been discussed for a success of participative design. The design activity is necessarily local (Wilkinson 2011), where the necessity to coexist with change is recognized (Folke et al. 2005). Immersed in gradual change, people may neglect, survive, or follow through. However, the rapid change of a disaster leaves no room to ignore the severe situation. Rather, mind-sets and status quos are challenged. People are asked how to take advantage of unprecedented changes, and turn crisis into opportunities for development (Carpenter and Gunderson 2001). This is not merely the reconstruction of broken roads and bridges but the rehabilitation of human-nature systems.

The rehabilitation of socio-ecological dynamics during periods of rapid change requires adaptive learning (Batterbury 2010) through temporal and spatial scales: (1) learn to live with change and uncertainty, (2) combine types of knowledge and learning, (3) create self-organization toward resilience, and (4) nurture resources of

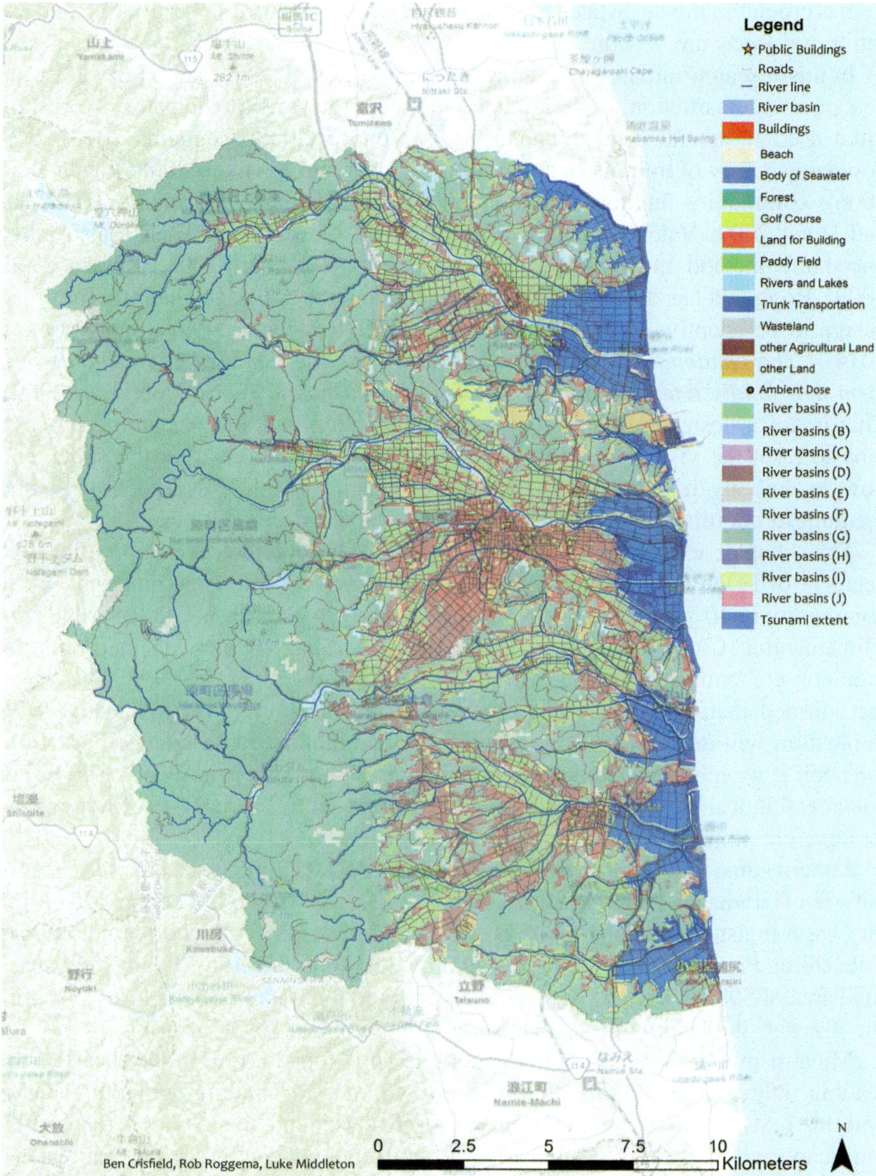

Fig. 6.2 Substantial flooding along the Japanese east coast as result of the tsunami (Author: Wanglin Yan)

resilience for renewal (Berkes and Folke 1992; Folke et al. 2003). However, what is social-ecological dynamics?, and 'how can a community set goals and measure progress toward them as they plan and design for future development?' (Girling et al. 2006). Locality shall be a key and it always brings challenges in practice.

This is crucial in the devastated area of Japan, where population is shrinking and the future prospects are obscure (Kabashima 2012; Hirano 2013).

In this research initiative design charrette was chosen as a method in landscape design of reconstruction. A design charrette is an intensive participatory workshop often used in spatial design (Glenn 1994; Condon 2008; Roggema 2013). It is popular as a way of training in classrooms (Walker and Seymour 2008; Eagen et al. 2006), community building (Condon 2008), and planning consultation (Howard and Hurst 2009; Valencia-Sandoval et al. 2010). Goedert and Tanaka also introduced the method in post-disaster reconstruction (Goedert 2008; Tanaka et al. 2009). Roggema has developed a theoretic framework of design charrette for urban design to be adaptive to gradual climate change (Roggema 2009; Roggema et al. 2014). "*The challenge of charrettes is to integrate the information and tools in a form that fluidly intersects the decision-making discourse*" (Condon et al. 2009). This is not a design technique but a communication and social learning process, how to reset the vision of the society in uncertain conditions after the shock and how to embody it into policy. Landscape design and plan is used as a spatial medium to coordinate governmental programs and citizen activities.

This chapter examines the severe impacts of the unprecedented radioactive accident of FDNPP on local community and created a sustainable blueprint of the community in 40 years by a design charrette. The project area is Ota River basin in Minamisoma City, Fukushima Prefecture of Japan, which is part of the radiation-contaminated region within 20–30 km sphere of FNDPP. The result reconfirmed that though the shock is severe and the impacts of radioactive contamination will drag for several decades, the community could bounce back from the nadir if we rebuild the industry and landscape based on the intrinsic resources of social-ecological systems. The term of 40 years is set to the half-life of radioactive isotope, caesium-137.

Minamisoma City was established on January 1st, 2006, as result of a merge between Haramachi Town, Kashima Town and Odaka Town. In February 2011, the city has a registered population of 71,561 people. Located in the northern corner of Fukushima Prefecture and facing the Atlantic Ocean to the east, Minamisoma has a total area of 398.50 km^2, 55% of which is covered by mountains and forests. Before the disaster, the leading industries were agriculture and manufacturing.

Minamisoma is located approximately 25 km^2 forests. Before the disaster, the leading industrial area, consisting of the town and surrounding agricultural fields was hit in two ways: the earthquake caused a tsunami that flooded the coastal landscape and it also hit the nuclear plant, causing a radioactive plume that reached the Minamisoma area. On the map (Fig. 6.3) one line represents the tsunami-flood (in purple) and the other circular lines (in yellow) the distance from Fukushima Nuclear plant, e.g. the radiation level. A radiation plume determines the activities that can be undertaken in areas with different radioactive levels. The government has decided to allow no building activities between the tsunami line and the coast. These two disasters at once had also other impacts. 70% of the farmers in the area

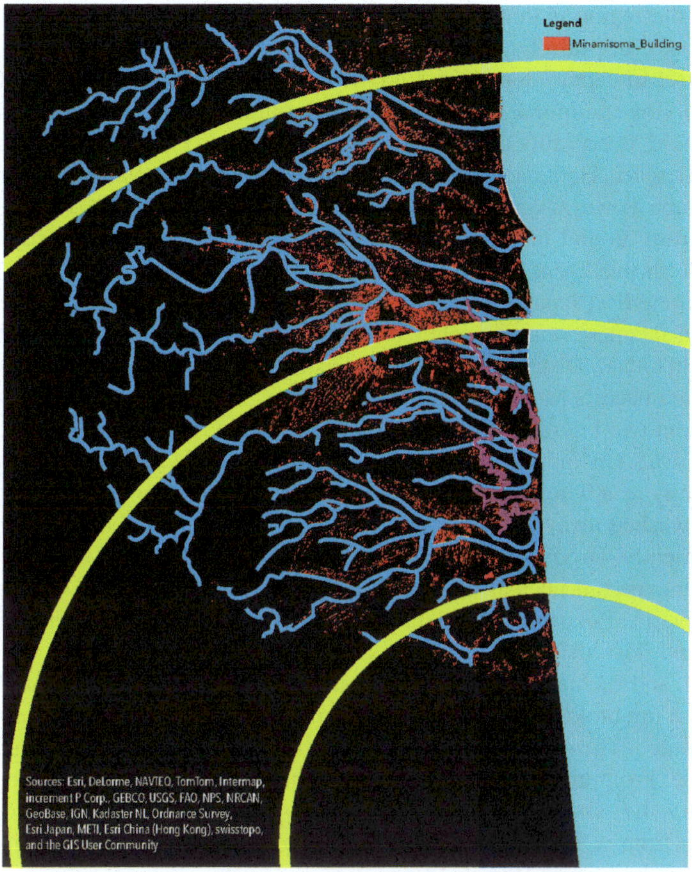

Fig. 6.3 The double disaster of nuclear contamination (the yellow circles determine the influence from the nuclear plant) and the purple indundation boundary determining the impact of the flood (*Source* Roggema 2015)

do not work anymore, and many young families have left the area. The risk of population and general decline lurks, especially when growth is concentrated in the valley where the Shinkansen rides.

6.2 Study Area

The study area, Minamisoma City of Fukushima Prefecture, is a new city created through the merger of the towns of Haramachi, Kashima and Odaka in 2006. The city is part of the Soso District of Fukushima's coastal Hamadori region, with a population of 70,878 in 2010 and an area of 398 km^2. The most distinguished feature of the region is that Minamisoma hosts the ‘Soma Noma-Oi’ (Soma Wild Horse Chase), deemed by the national government a ‘significant intangible folk cultural asset of Japan’ due to its 1200-year history. This is an exciting three-day event involving traditional ceremonies and samurai horsemen clad in armour and helmets, held every July in the Soso region (Soma City, Minamisoma City’s Kashima District, Haramachi District, Odaka District), and Futaba region (Namie Town, Futaba Town, Okuma Town), with participation of up to 500 horses in total, gathered from around the country. The Ota Shrine in Haramachi District plays a central role in this event.

In the Great East Japan Earthquake of March 11, 2011, an earthquake registering nearly 6 on Japanese seismic scale (comparable with Richter scale from 1 to 6) was measured in Odaka District and Kashima District, and nearly 5 on Japanese seismic scale in Haramachi’s Moto-cho and Mishima-cho. The tsunami subsequently swept an area about two kilometres inland from the coastline. Inundation damage affected an area of 15.8 km^2 in Kashima, 10.5 km^2 in Odaka, and 14.5 km^2 in Haramachi, while the city as a whole counted 631 casualties, seven missing persons, and 2875 buildings washed away.

Subsequently, survivors of the earthquake who had escaped from the tsunami faced an unexpected nuclear accident. Ten kilometres south of the city, in Futaba region (Fig. 1), Fukushima Daiichi Nuclear Power Plant was damaged by the tsunami and lost its main electrical power. Hydrogen explosions occurred in Reactors 1 and 3, fuel rods were exposed in Reactor 2 due to a lack of cooling water, and a fire broke out in Reactor 4 and released a large amount of radiation into the atmosphere. The overall accident was rated Level 7 on the International Nuclear Event Scale. From immediately after the earthquake and tsunami on March 11 until March 16, the situation grew increasingly serious, with evacuation orders expanding from a radius of 2 to 10, 20, and then 30 km, throwing the local communities into chaos (Fig. 6.4).

As a result, the city of Minamisoma was split by the tsunami in the coastal area and, further inland, by 10-, 20-, and 30-kilometer exclusion zones.

The local impacts of the nuclear accident are severe. The total population of Minamisoma City dropped from 70,878 before the accident to 62,397 at present of August 1, 2016. The major reasons why people left were concerns about radioactive contamination, and loss of employment. Previously, this city had flourishing agricultural and forestry industries. Thousands of Haramachi’s households were either part- or full-time farmers. High concentrations of radioactive contamination accumulated in the forests and soils of the Abukuma Highlands running north to south in the western part of the city’s territory. This area is designated a part of an exclusion zone (occupancy prohibited). Water flows out of this area via the Niida and Ota Rivers, before flowing into Pacific Ocean. As a result of the situation, paddy fields irrigated by this water are repeatedly exposed to risks of soil contamination. Despite painstaking farmer-driven efforts at decontamination, the area of rice planted plummeted from 25,900 hectares before the disaster to just 750 hectares in 2016 (“Crop statistics,” Ministry of Agriculture, Forestry and Fisheries).

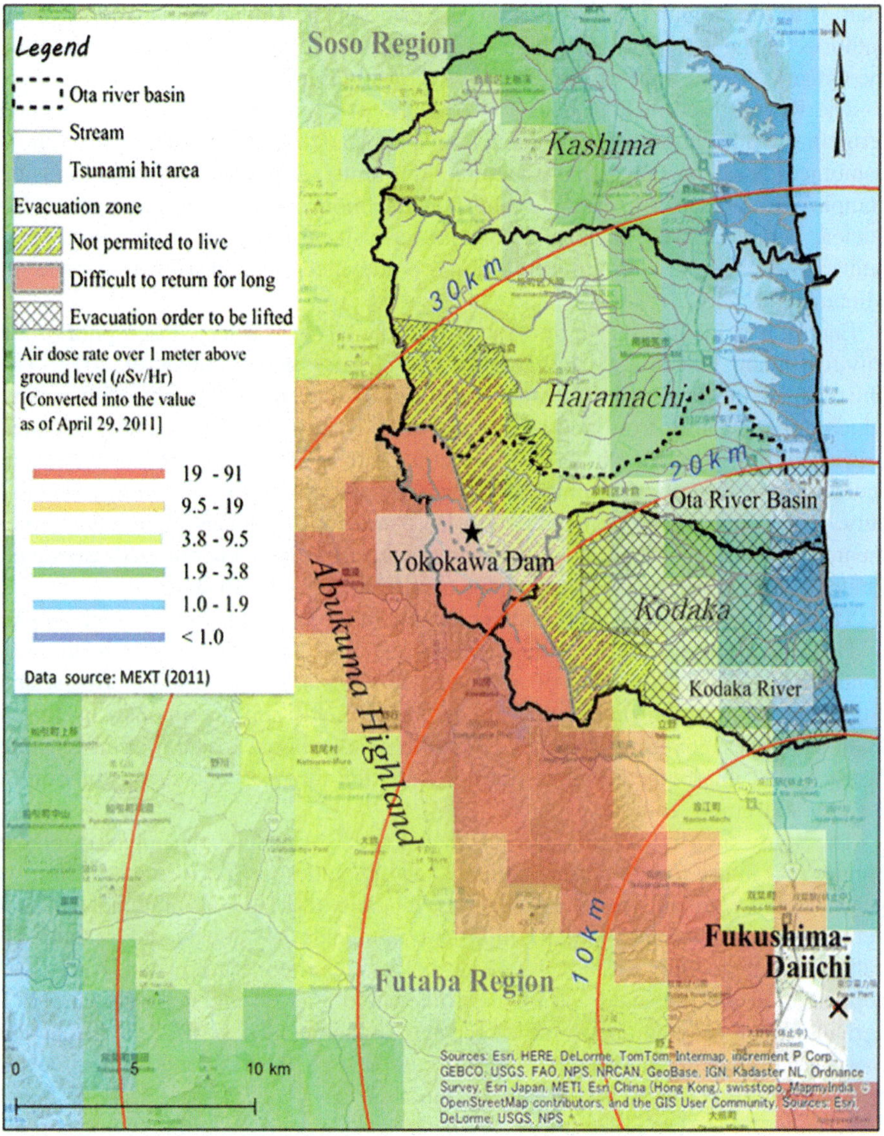

Fig. 6.4 Locations and zones of Minamisoma City (Ota River basin is indicated by bold line) (Author: Wanglin Yan)

Consumers still stay away from Fukushima's food and agricultural products. The produced rice is mainly provided for husbandry and industrial use.

The city began reconstruction efforts promptly after the emergency situation from the nuclear accident eventually stabilized. In August 2011, the Minamisoma City Reconstruction Vision was adopted, and in December that year, the Reconstruction

Plan, with the slogan "Reconstruction, the Pride of the United Minamisoma," was authorized. The Reconstruction Plan was adopted based on extensive discussions by the Citizens' Committee of the City for Reconstruction.

The Reconstruction Plan calls for a review of regional disaster risk reduction programs, with earthquakes, tsunamis, and nuclear disasters in mind, as well as combined disasters. For tsunami victims, 350 units are being built in ten housing complexes as disaster-resistant public housing more than 20 km away from the nuclear plant, and 303 households of disaster-resistant housing clusters are being built in 21 sites for group relocation on elevated ground (from website of Minamisoma City).

Energy was identified as a priority sector. The city was selected as a pilot of the Environmental Future City Initiative of the country in December of 2012. The objective of the project is for the community to live in coexistence with nature, which implies a shift from nuclear to renewable energy. The municipality declared no nuclear policy on March 25, 2015, ensuring citizens free from the fear of nuclear accidents and bringing back hopes for next generations to live in the city sustainably. Renewable energy projects by solar panels and biomass plants are implemented, and energy reduction and conservation actions were encouraged. However, fundamental energy shift is waiting for transformation of industry.

The design area of this workshop is located in the basin of the Ota River, a part of Haramachi town before a municipal merger in 2006. This district in the city contains evacuation divisions of non-entry zones, planned evacuation zones, and emergency evacuation preparation zones. Paddy fields and greenhouses were the main land-uses covering the area before the Fukushima accident. The ensuing disasters and intertwined zoning systems made the recovery and reconstruction of the area extremely complex.

The study area is close to the urban centre of Minamisoma and was for the biggest part in use for agriculture before 3.11. It includes an important shrine for the yearly horse race and festival, which is officially declared as national cultural heritage. The landscape with its hills and farmlands are considered as national cultural heritage. Only 20% of the paddy fields are currently cultivated. Aging and depopulation are the major problems. Renewable energy, green houses and ICT agriculture are seen as opportunities for new industries. These new land uses and new agricultural business models are discussed with the residents. These developments require a landscape design for the area.

A landscape design process was conducted using a design charrette workshop for the Ota area in Minamisoma City. The Ota area is an agricultural belt in the southern portion in Haramachi district of Minamisoma City. The Ota River originates in the Abukuma Highlands, passing through the Yokokawa Dam reservoir as it flows from west to east. The belt is an expanse of rural landscapes with small groves near traditional houses and paddy fields reminiscent of the historic town of Nanto (Toyama Prefecture, about 500 km away) with its dispersed settlements (Suzuki 2013). Historically, the Soma Nakamura domain accepted many migrants from what today is Nanto in Japan's Hokuriku region, to help the population recover from significant losses in the Tenmei Famine (1783) (Okada and Kurosu 1998).

In the middle of the study area is the Ota Shrine, which made a considerable effort to revive the Soma Noma-Oi. Further upstream is the Baji Koen (equestrian park), with a lane for horse-riding lessons and competitions, although today visitors are shocked to see it has become a landfill site for decontaminated soil. Downstream is the coastal area of Haramachi District, which was devastated by the tsunami.

6.3 Issues and Challenges

The question after every disaster is whether the lost buildings and infrastructure will be rebuild. In many cases the population desires their properties to be rebuild in the same locations and in similar ways. However, this is not always the best strategy as these new houses are, from the moment they are realised, evenly vulnerable as they have been before the disaster. Alternatively, new buildings can also be located where a disaster cannot harm them in the future. Or, if no other area is available, the buildings should be adjusted to and prepared for the impact of future disasters. In many cases, when disasters are unpredictable, this is very difficult.

Many reconstruction projects have been carried out and no one has the one and only solution. What is known is that technical solutions tend to end in a debate between proponents and opponents. This in itself causes delay, mere discussion and disinterest of the residents. For the Minamisoma area a lot of research has been carried out on the radiation levels (Cyranoski and Brumfiel 2011; Furutani et al. 2012), the water quality (Endo et al. 2013), demographics (Zhang et al. 2014) and social impacts (Campbell 2013).

Despite all this research the key problem is merely addressed: how can the people living or returning to the area become proud on their land again. Proud on their heritage, their families and their lives. The question of recovery is how the people can be reborn. This is only possible to rethink what the future needs to be and subsequently hope the people will return to their homeland.

6.4 Challenges in Reconstruction

6.4.1 *Reconstruction Principles*

The Reconstruction Plan declared basic principles such as livelihood restoration, economic reconstruction, safety and security, and self-help, mutual help and public (government) help. However, there is a large gap between objectives and practices. Many objectives address long-standing challenges that existed even before the disaster and could not be easily resolved conventionally. The government has to tackle these challenges, such as economic revitalization, maintaining population numbers, and improving social services. When nuclear contamination is added to this mix, situation becomes extremely difficult. Topics such as harmony with natural environment and sustainable development are buzzword, so they do not articulate any unique features of this region. Citizen-based mutual help and collective help are popularly spotlighted, but lack concrete measures. Rather, these concepts are criticized as pretexts for the irresponsibility of government. In actuality, the idea of *creative reconstruction* is no more than *recovery* from the problems that existed before, and it is rare to see organizational efforts of turning reconstruction into opportunity for transformation, from a historic and holistic perspective of planning and design (Ishikawa 2013, 2014).

6.4.2 Community Devastation

Due to the tsunami and nuclear disaster, Minamisoma City was divided into a puzzle (Fig. 6.4). Based on the concentrations of radioactive contamination, the city was split into zones: one where it will be difficult for residents to return for a long time, one where residents are not permitted to live, and one where evacuation orders are ready to be lifted. Also, a tsunami hazard zone stretches along the coastline. As a result, one block could have completely different situation with compensation and subsidies compared to another across a street. Such differences make it difficult for residents to get together, not to mention making consensus.

During the period immediately after the disaster, *bonding* and a sense of solidarity was popular all over the region. This sense has been described as the period of disaster utopia (Solnit 2010) when people share the same circumstances and have the same issues in mind. However, after the situation was getting tranquil, people tend to return their attention to personal life, and communication declines.

As the period of living in temporary housing dragged on, people started to become irritated (Shineha and Tanaka 2014). Among the total death of 1121 persons by the disaster 485 persons were recognized as indirect deaths due to the shock during evacuation (from website of Minamisoma City, July 6, 2016), the top rate in 33 tsunami-stricken municipalities. Decontamination progressed slowly, rumors on food safety abounded, and employment declined. Concerns mounted about future employment and health. When residents are dispersed, a community could cease to exist, and culture and traditions could disappear (Samuels 2013). These human and cultural dimensions rarely make it onto the government's radar. The citizens may be keen to act, but it is difficult for an individual to take up leadership and undertake actions. Municipal servants, key human resources in reconstruction, are in severe situation too. About 15 to 20% of servants in municipalities within 20 km of FNDDP has mental problems, 10% of them has applied early retirement.

6.4.3 The Struggle Against Nuclear Contamination

The slow pace of decontamination became an enormous obstacle for reconstruction. The residents ardently desired to return to normal life, first step of resilience. However, with polluted farmland, lost employment, and an aging population, the wish would not be easily realized. It is possible to reduce radioactive cesium (cesium-137) concentrations to one-fifth the levels in controlled paddies by removing top soil (Sakai et al. 2014). As of June 30, 2015, however, the ratio of the area decontaminated was only 19% for residential land, 13% for farmland, 45% for forests, and 6% for roads (from website of Ministry of the Environment). The community wondered if even after decontamination, the forest, water, and soil would be truly safe. Because of inadequate responses at the time of the nuclear accident, residents developed a deep-seated distrust of bureaucrats, the government, and researchers (Zhang et al. 2014). Governmental systems are also perceived to be rigid. For instance, rice planting is not permitted on the paddy fields that were contaminated. However, by regulation no conversion to other land uses is permitted. Meanwhile, farmers strongly wish to restore the polluted land to use as paddy fields, amid the concerns about a shrinking labour force and finances.

6.4.4 Social Resilience

Therefore, the disaster impacts are huge and the challenges are complex. Local communities were prepared to act and contribute to reconstruction, in contrast to the slower actions of the government. A total of 169 non-profit organizations (NPOs) are registered in the city, the majority established after the disaster, and they are active in a variety of fields, including culture, arts, sports, social services, support for persons with disabilities, children, education, international cooperation, environment, and urban development. As the identity of the area, the Soma Noma-Oi horse festival event had to be held outside the exclusion zone in July 2011, but starting in 2012, it returned to the traditional locations. Some farmers, unable to resume rice growing immediately, have been busy finding alternative businesses, including projects to grow canola and produce biodiesel, to construct a mega-solar (large photovoltaic) plant, and "solar sharing" projects (farmland being used for both growing crops and for photovoltaic power generation) with the support of NPOs and local companies. Although the activities are mostly non-profit and the effects are small, comparing with government led infrastructure projects, they own the power to change the society from large-scale, centralized industry to community-based distributive systems. The issue is to restore the energy of the community in a way its members would like to support incrementally for a better future of their area and place each small piece into a whole picture. Here is the necessity of a spatial plan to bridge the gaps among stakeholders and revitalize the potential powers of local resources. A design charrette is a valuable way to involve people, to give them the lead in thinking about the future, and to come up with a plan that they support and want to realize together. The government holds the role to support the community and make the activities contribute to the realisation of the Reconstruction Plan.

6.5 Design Principles

After the triple disaster (earthquake, tsunami and nuclear explosion) struck Japan, the affected area is undergoing recovery, but the reconstruction in the radiation-contaminated area faces more hindrance than the areas that were only struck by the tsunami. Several issues should be taken into account:

1. Decontamination is a prerequisite of reconstruction, but this progress is slow in areas where temporary storage sites have not been decided on. It is also turning out to be quite difficult and time consuming to acquire the necessary approvals from landowners.
2. Depopulation: An outflow of migrants after the Fukushima accident has lead to an ageing and depopulation problem in Minamisoma. The recovery of the city, even in areas proven to be low radioactive, is hindered.
3. Mistrust: Because of the inadequate pre-disaster precaution and ineffective post-disaster treatment, the public mistrust on the government, TEPCO and the government-funded scientific institutions influence the recovery of Minamisoma.
4. Land-use: As the soil was contaminated in the Fukushima accident, swaths of farmland in Minamisoma could not be farmed anymore. In addition, the seaside area that was affected by the tsunami could not be used as a residential area as it used to be. How the contaminated land can be re-planned is a vital issue in Minamisoma.
5. Old industries of Minamisoma, agriculture and manufacturing declined rapidly after the accident. New industries, such as new energy and bio-agriculture, are the main engines of the city's recovery.
6. Community is often regarded as a vital factor in the reconstruction after a disaster. The question is if and how the communities have changed after the Fukushima accident in Minamisoma. Are communities still co-operating and functioning during the reconstruction process?
7. Lack of leadership is a main cause of the long time aftermath in Minamisoma. Could new leadership in the communities be developed that accelerates the recovery?

6.6 Aims of the Design Charrette

Several overarching aims for the design charrette have been defined:

1. To understand the people (residents and outsiders) and the area;
2. To collaborate with the people (residents and visitors);
3. To design a vision and a plan for the village;
4. To design the pathways and spatial alignments for the future;
5. To make recommendations for the implementation of the design.

These overarching aims have further been detailed in the program of demands for the design charrette. The following points have been considered during the workshop:

- The purpose of the workshop is to support the reconstruction process. The main issues are how people could be supported in recovering from the disaster, and how agriculture can be reconstructed, given the long-term problem of soil contamination, in a shrinking and ageing community.
- Use of global sustainability goals when rebuilding the town of Minamisoma, such as food security, energy security with renewables, a bonding, bridging and networking community, a de-growth economy in a shrinking and ageing society, a green economy with accompanying resources, conservation of cultural heritage and habits and business innovation.
- Look at the long-term future and take into account the visions and policies at global and Japanese levels. The plan should transcend the disaster and overcome the shrinking trend, as the village is predicted to have disappeared in 2040 according to the population prediction.
- Use the features of the Ota district. Transform the negative characteristics, such as low population density, relative conservative mind-set and the orientation on agricultural industries. At the same time use the positive drivers, such as the strong motivation to shift to a community based agriculture, new activated ties with NPO's from outside the area, several big projects supported by the state in the field of renewable energy, ICT-farming and decontamination, in the area to overcome the negatives and create a prospective view for the future.

6.7 Pre Study—Analogues

Given the context of the Minamisoma area and the vulnerability of its coastline for the possible impacts of a tsunami, a pre-study has been carried out on the possible visionary responses that may be possible. This study consisted of the application of the design propositions made for Rising Currents competition and Exhibition in New York (Bergdoll 2011), after the hurricane Sandy impacted on the city, to the landscape of Minamisoma. Five visions have been redesigned to the Minimosoma area.

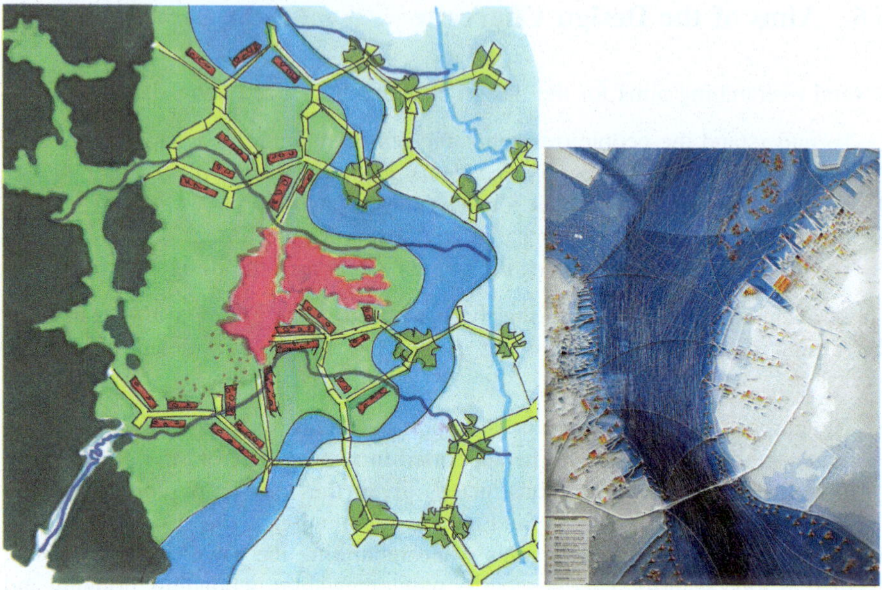

Fig. 6.5 Aquatius City

The Aquatius city (Fig. 6.5), designed by nArchitects, is 'a form of urbanism in which the city extends into the water, and water enters the city. An ever-growing population is accommodated by aqueous neighbourhoods featuring suspended housing, wave-attenuating piers, and service by a new generation of biogas ferries and a tramway. Land neighbourhoods are punctuated by basins and culverts that absorb storm runoff and function as parks in dry weather, designed by nArchitects.

New Urban Ground (Fig. 6.6), designed by ARO and dlandstudio, proposes 'a landscape of absorbing material for the highly polluted Gowanus Canal and rethinks comprehensively the texture of the coastline and the urban street. Transformed through wetlands the coast is a complex natural instrument for filtering both tidal change and street runoff after storms. The streets themselves are 'greened' through surfacing in absorptive, open-mesh concrete tiles and a layered filtering system' (Bergdoll 2011).

Oyster Tecture (Fig. 6.7), designed by SCAPE, is a proposal for 'a massive wildlife restoration. Oysters are transforming the polluted waterway into a very productive oyster hatchery, while the mollusks assist in cleaning the water. After being hatched the oysters are relocated to create a wave attenuating reef and a new natural aquatic park, protect the adjacent shoreline and stimulate the growth of other marine life. Planning is not a static design but implementing and exploitation of dynamic natural systems' (Bergdoll 2011).

In Waterproofing Ground (Fig. 6.8), designed by LTL Architects, 'the existing land is cut and redeployed to achieve a varied topography an the flat site,

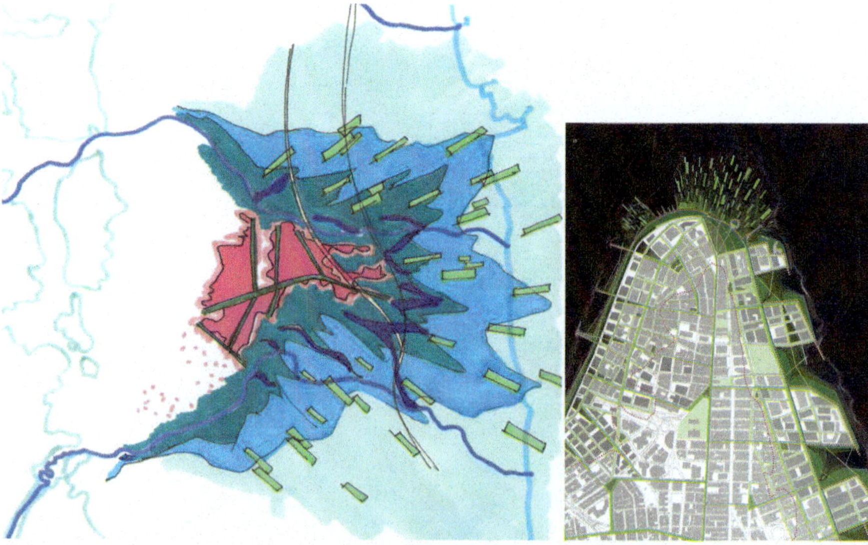

Fig. 6.6 New urban ground

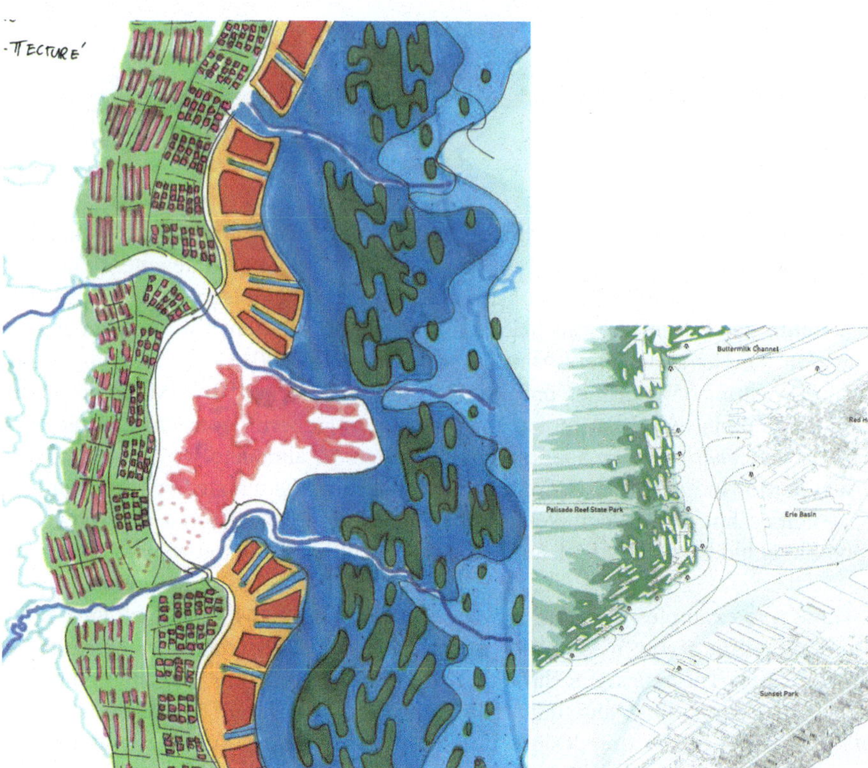

Fig. 6.7 Oyster tecture

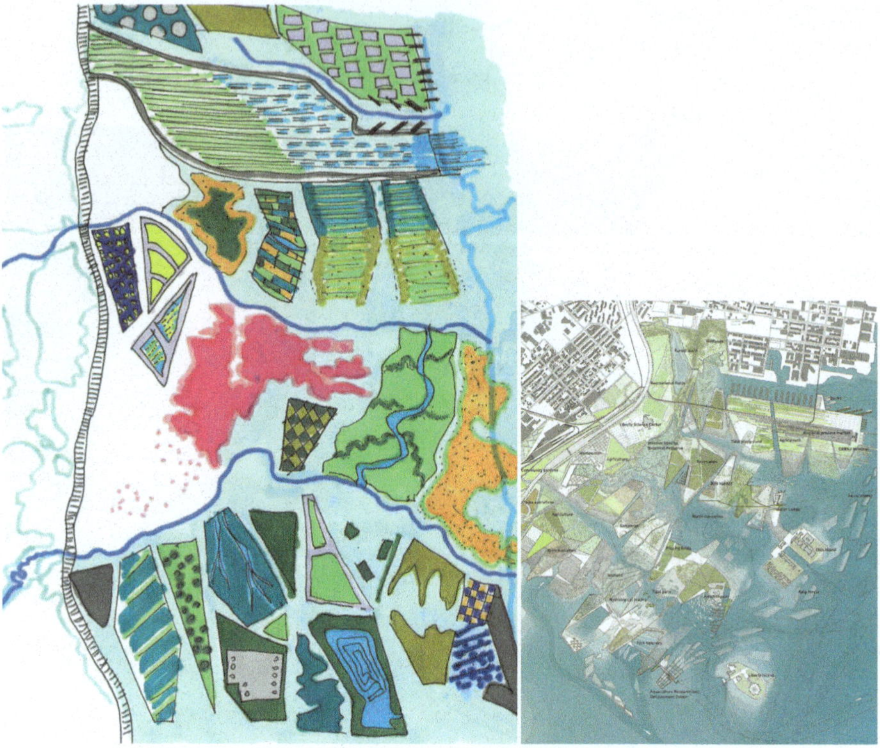

Fig. 6.8 Waterproofing ground

a crenelated landscape of jagged fingers that by lengthening the coastline manifold allows it to attenuate waves and serve as a natural filter of tidewater. The new topography is zoned for a range of urban functions, from leisure and recreation to aquaculture and commerce, all placed on the site in relationship to the degree of protection the elevation offers from flooding' (Bergdoll 2011).

The Working Waterline project (Fig. 6.9), designed by Matthew Baird Architects, emphasises 'the implementation of a large berm, consisting of recycled glass 'jacks', distributed in shallow coastal waters to attenuate waves, particularly during storm surges. The berm and the aquatic reefs create new recreational opportunities but do not completely transform the site's industrial character. The proposal includes new natural and economic ecologies for the region' (Bergdoll 2011).

These five examples illustrate how the coastline in the Minamisoma area could be redesigned in a way it improves the safety and at the same time increases the spatial quality. The redesigns show that a dissipative structure may strengthen the coastal protection zone and attenuates the wave activity. The different design options were used in the designs for both Minamisoma and Kesennuma (see Chap. 7).

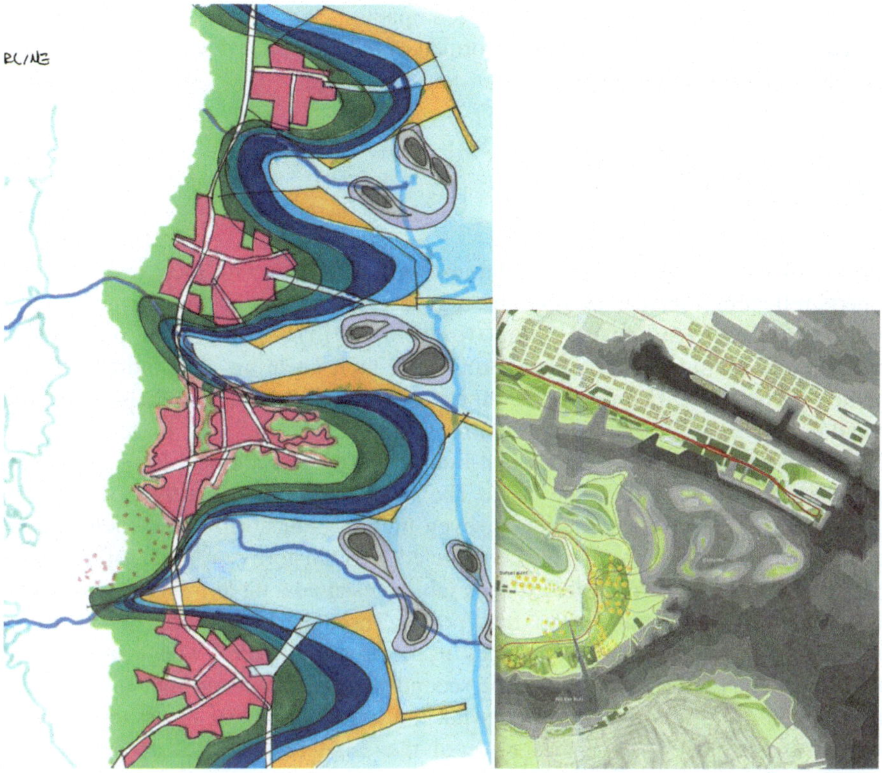

Fig. 6.9 Working waterline

6.8 The Design Charrette

A charrette workshop is a method used often in urban planning around the world (Condon 2008; Girling et al. 2006; Roggema 2013), though not very common in Japan yet. The charrette workshop was held over four consecutive days, from September 8 to 11, 2014. Prior to the workshop, notifications were sent to NPOs, Minami Soma Decontamination Laboratory, Ota Reconstruction Committee, and municipal council members, as well as teachers and students from the local Soma Agricultural High School, and full-time farmers. They were invited to participate in a '30-30 Exercise' on September 9, and a participative design workshop on September 11. A set of geographic data and documents about the area was prepared beforehand for experts and carefully reviewed in the morning of the first day. Outside experts and participants, guided by locals who are familiar with the area, walked around the subject area, and collected information as input for the workshop. These activities included visits to devastated buildings, temporary housing, forests, farmland, agricultural companies, and cultural spots. They also visited the city's museum and were informed of the history of regional development and the Soma Noma-Oi festival.

6.8.1 30-30 Exercise

The 30-30 Exercise is essentially a brainstorming session with local residents and experts reflecting together on how the region was 30 years in the past and how it could become 30 years in the future (Roggema 2013). A total of 28 participants broke into four groups and gathered around the tables bearing signs indicating four themes: Population, Economics, Culture, and Environment. They discussed the situation 30 years ago, wrote key words on "sticky notes," and placed them on aerial photographs. Several minutes of brainstorming was allowed for each theme, with the post-it notes left on the maps. The participants subsequently moved to the next table and added ideas during a seven-minute brainstorm about their second topic. Eventually, all participants conducted a brainstorm on each of the four themes. Finally, the team leaders at each table presented a three-minute summary at each table of the results up to that point. Through this process, each group reconfirmed things such as what existed in the area in the past, what was lost, and what should be kept for the future. The second round of brainstorming followed the same structure but the question was what the future should or could look like in 30 years future. The participants covered the same four themes as before, and summarized the results of each theme in three-minute presentations at the end. The most significant or the "hottest" topics for 30 years ago and 30 years in the future are summarized in Table 6.1.

The output of the 30 × 30 Exercise is full of messages, as summarized in Table 2. Residents felt proud about elements such as the rich local environmental setting, their history and culture, for instance the Soma Noma-Oi horse festival, and bonds within the community and family, which they had strongly experienced through the disaster. On one hand, compared to 30 years ago, population had decreased, there were fewer young people, and the impacts of the earthquake, tsunami, and nuclear accident were severe, but on the other hand, people showed a tendency to see a bright future. A radical change in thinking is required to leave behind the dark image of the disaster. Residents put forth many ideas, including the restoration of local culture, promotion of local self-sufficiency, the use of renewable energy, and the promotion of collective farming. It is worth mentioning that many participants clearly remembered the opening of the Fukushima Nuclear Power Plant 30 years ago; not surprisingly, none of them took it as a positive thing during the workshop. Instead, they hoped that the radioactive contamination would be gone by 2040 and that all homes could be equipped with solar power. This reflected a striking change in people's mind-sets after the disaster, and provided common ground for a transformational landscape 30 years in the future.

Table 6.1 Hot topics of the 30-30 workshop

30 years ago	30 years in future
Population	
Bigger population than today	Automated hospitals
Population was balanced	Compact cities are possible
Large families were normal	Elderly able to care for each other
Many were farmers	Super-aged society, average age 95
Many were living/working away from home	More farmers due to agricultural renovation
Communities active in the region	World-class health facilities/operations for elderly
Economy	
Economy was growing	Self-sufficiency put into practice
Daily life was vibrant	Almost 100% of homes have solar power
High quality of industrial products appeared	People produce electricity by themselves
Entrepreneurial spirit	Gasoline is 1000 yen per liter
Transportation was convenient	Other construction materials have replaced concrete
Greed was rampant	Possible to live on farming
Culture	
Overseas travel became popular	Noma-Oi traditional horse festival continues
Aum Shinrikyo cult group and subway sarin incidents	Famous worldwide as a horse town
Nuclear power was operating	City returns to conditions before the disaster
Cars, telephones	Regular international events on theme of horse races
Super Nintendo Entertainment System	Noma-Oi traditional horse festival recognized as World Heritage
Video tape	More new events
Environment	
Increase in pollen allergies	Fukushima Daiichi nuclear accident has been dealt with
Serious pollution of urban rivers	CO_2 emissions gradually declining
Weak regulation of agricultural chemicals	Effects of radioactive materials have disappeared
No air conditioners at home	Global warming continues
Rampant resort development	Renewable energy is a part of everyday life
Standard of living was rising	An increase in various kinds of disasters

6.8.2 Design Concepts

During the site visit and communication with local participants, organizers were strongly impressed by the popular narrative that Minamisoma City is closely connected with the philosophy known locally as Goshiho, which was articulated by Sontoku Ninomiya (a prominent 19th-century Japanese agricultural leader and philosopher, 1787–1856). The basic principles of this Goshiho philosophy are *shisei* (sincerity), *kinro* (work), *bundo* (proportion), and *suijo* (yielding). *Bundo* means to determine the appropriate maximum that should be paid for something, while *suijo* means to prepare or save for the future (Armstrong 1912). It also means to share a part of your income with other people. It is said that this philosophy helped the region to recover from the Great Tenmei Famine (1781–1788) and Great Tenpo Famine (1830–1844) (Okada and Kurosu 1998). The principles later lost influence in the process that abolished clan domains and replaced them with today's prefectures, but the teachings of Sontoku are still alive in this region. Goshiho is actually part of the Minamisoma City Citizens' Charter.

The Great East Japan Earthquake of 2011 was the biggest disaster in this region since the Tenmei and Tenpo famines. This study's authors believe that Sontoku's teachings provide important community DNA "genetics" that could inspire and guide efforts to overcome the crisis. Rather than to base the future on the consumerism of the past thirty years, a sustainable society is about living lives that are compatible with one's capacity, are embedded in the natural landscape, boost productivity through solidarity, and prepare for the future.

In Minamisoma, the Soma Noma-Oi horse festival has been a continuous cultural symbol for the past 1200 years, and it is considered to be precious to the community, even after the disaster reconstruction. People are making an effort to pass on the tradition to future generations. It is another source of inspiration for reconstruction.

Local people have moved quickly. For instance, considering the fact that it is impossible to reopen many paddy fields, and the construction of new dwellings is prohibited in areas that were inundated by the tsunami, the people came upon the need for alternative crop plans that match the land conditions, rather than solely rice crops. The design charrette process reconfirmed the motivation of the community to quit small-scale farming and make a shift toward collective production. The aim of the workshop was to work with participants to realign hitherto fragmented programs, projects and activities into a blueprint for a sustainable future in the area.

6.8.3 Creating the Design

In the workshop design process a series of actions was determined:

1. Identify the key spatial elements of the region
2. Set the time frame
3. Determine land use policies
4. Configure a spatial landscape framework
5. Locate spatial elements in the landscape
6. Undertake trial calculations for the supply-demand balance of food and energy.

Key spatial elements identified in the region included the Ota River, Ota Shrine, as well as elements related to the Soma Noma-Oi horse festival and derived from the Sontoku principles. Considering the half-life of the radioactive materials, mainly cesium-137 (Nakano and Povinec 2012; Rossi et al. 2013), the target year was set at 2040.

6.9 The Design for Minamisoma

During the charrette it became clear that solutions of a technical nature only would not suffice. The mechanical approach of building strong seawalls is causing debate between proponents and opponents, but doesn't deliver what is needed to help the local population. Therefore the desires for a future are written down and grouped (see Chap. 5) and made the context clear. The identified problems and the wishes of the local population have led to an approach that meets the basic questions.

The people in this area lost their pride. That is the reason why so many left. The main design task is therefore how to give the people their pride back. How can they become proud again on their own land, their heritage, their families and their lives. For sure, this isn't going to happen with a stronger seawall and other technocratic solutions to deal with the impacts of the disasters that happened. This has to come from somewhere else. People must be reborn. Therefore we started this plan from a philosophical basis.

The thinking of Ninomiyasongtoku takes nature as its core. Live with nature and build with (the principles of) nature. Learn from nature and give back to nature. Create self-sustaining places where cycles are closed. The understanding of the landscape and appreciation of its heritage allows us to create embedded landscapes (Fig. 6.10).

In Minamisoma two genetic characteristics shape the basis for the future: the horse festival and the ideal landscape.

6.9.1 Horse Festival Ceremonial Route

The horse festival is a very old historic event and connects heritage symbols in the landscape through a heritage route (red lines in Fig. 6.9). Every year, 400 horses come together in Minamisoma during this three-day festival. This festive route is unused for the other 362 days of the year. To increase the impact and meaning of the festival the route is used to create a year-round calendar of events to give rebirth to the impact and meaning of the horse festival.

In the early days 200 horses were born, raised and kept in Minamisoma prefecture, but nowadays this number is only 20. The horse keeping of 200 horses is

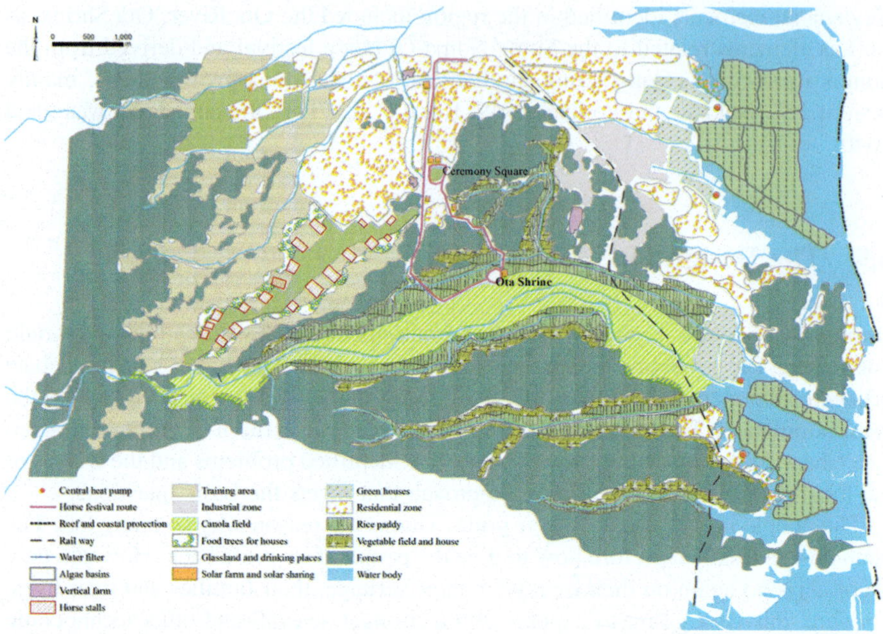

Fig. 6.10 Regional design for the Minamisoma area (Roggema, Yan, Middleton and Oba)

proposed to be re-located in the area. These horses are kept, groomed, given food, trained and can be seen in the vicinity of Minamisoma, where they historically used to be, along a new waterway with drinking places. This area is connected to the festival route.

6.9.2 *Staged Restoration of Food Production in the Ideal Landscape*

The ideal landscape used to be everywhere in the landscape, but due to the radioactive plume most of this landscape is no longer productive. The spatial principle is the sequence of landscape elements from the basis of the valley. First the stream itself, then the rice paddy, followed by vegetable gardens and houses at the edge of the mountain/hill and finally the forest on top of the hill. The ideal landscape has lost its value. The beauty is jeopardised, the productivity is extremely low, hardly anyone working or living in this landscape. After the disaster this ideal landscape must also be reborn. The spatial plan takes these values as starting point and uses them to design a future. When every design decision is linked to these principles a coherent landscape will emerge. This is sustainability. The plan for the region rethinks every part of the landscape. The time frame for realisation is set at 40 years. The landscape is recreated according to four phases of cleaning up the water system. The first phase, closest to the City of Minamisoma is relatively cleanest. It can be expected that within two years food can be produced again. In the second phase, between two and five years, agricultural production can be restarted in three streams that will become available after radiation dropped. In these streams a filter is proposed upstream to ensure the quality is good enough for food production. Third phase consists of a similar plan for one side-stream, which is expected to be ready for agriculture between five and eight years. The central valley around the Ota-river is the area, which is set to last longest before it can be used for food production. It cannot be used for food production for a period up to ten years, but in the meantime it can be used for canola plantation, a plant that is independent of radiation and could deliver rapeseed oil. Moreover, the bright yellow flower provides a beautiful coloured landscape in the valley, surrounded by dark green hills.

The calculation (Tables 6.2 and 6.3) shows nearly enough food (rice and vegetables only) can be provided to supply the population of 42,000 people. There is a

Table 6.2 Productivity and demand of food

	Units	Rice	Vegetables
Productivity	kg/year/ha	5390	6780
Consumption	kg/per/year	60	92
Population	42,000 persons		
Land required	Hectares	420	570

Table 6.3 Phasing of the regeneration of food productivity

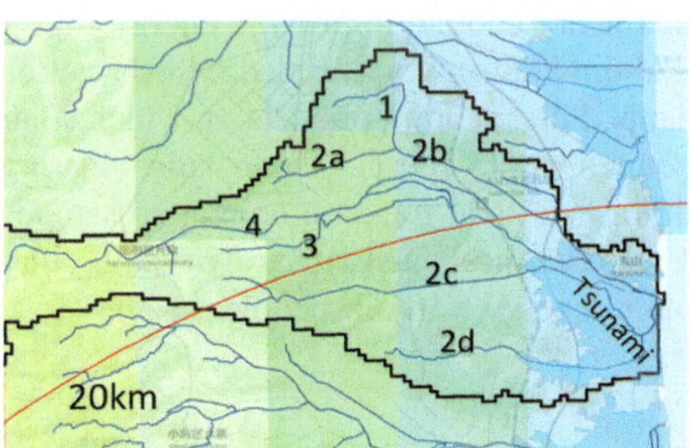

	Total area (ha)	Rice (persons fed)	Vegetables (persons fed)
Allocation of Land	100%	66%	34%
Phase 1	84	4980	2105
Phase 2a	150	13,870	5863
Phase 2b	112	20,514	8670
Phase 2c	176	30,949	13,080
Phase 2d	50	33,941	14,332
Phase 3	134	41,859	17,690
Phase 4	642	Canola	

small additional space required, which can be found in three vertical farm projects and in projected greenhouses near the algae fields.

6.9.3 Algae Fields

The tsunami line shows the flood impact of the tsunami. The government declared this zone to be kept free of any building. The government realises currently a protective wall at the coastline, but a similar tsunami still wipes out the wall and flood will enter the hinterland. Therefore an alternative protective system, attenuating wave energy, is proposed. Instead of a closed wall, a reef could decrease the power of the water once it hits the reef structure. The planned sea wall can act as such a reef, if it is perforated with little holes to slow down the water before it is let in the hinterland, where it comes to rest.

This water comes available to use in algae basins, where algae transform and clean the water into biofuel, meanwhile capturing carbon (twice their own weight). Moreover, algae operate as a bio-production plant, as they transform water (and air) into the biofuel, which can be used in CHP's to generate a combination of heat and power. The heat can be used in greenhouses to grow extra food and surpluses can be used in nearby residential areas. The power can directly be used in residential housing. The waste from the algae can be used to feed the horses.

The basins themselves add an extra layer in defending the land from a flood. The location of the basins leaves space for the existing monument for tsunami victims. The shrine forms an island in front of the basins and can be reached by a bridge, which makes the monument even more dramatic and scenic.

The energy generated in the algae fields is not enough to provide energy for all the 42,000 people in the area. There is additional electricity needed, which can be harvested through to extra solar power plants. Solar sharing projects, in which solar panels are put over agricultural crops, are a good option. These solar sharing projects are connected to the ceremonial horse festival route, at crucial crossroads and junctions, such as near the festival stadium/racecourse. The algae and solar sharing projects together generate enough energy for all people in the area. Surpluses are sold to neighbouring areas.

6.10 Alternative Model

In an alternative model the design teams explored how to implement a mosaic of function in the area. This patchwork consists of small entities of Japanese gardens, timber forests, solar sharing projects, rice paddy's and market gardens (Fig. 6.11). The rice paddy's and market gardens require intensive maintenance, while the other uses require less intensive care.

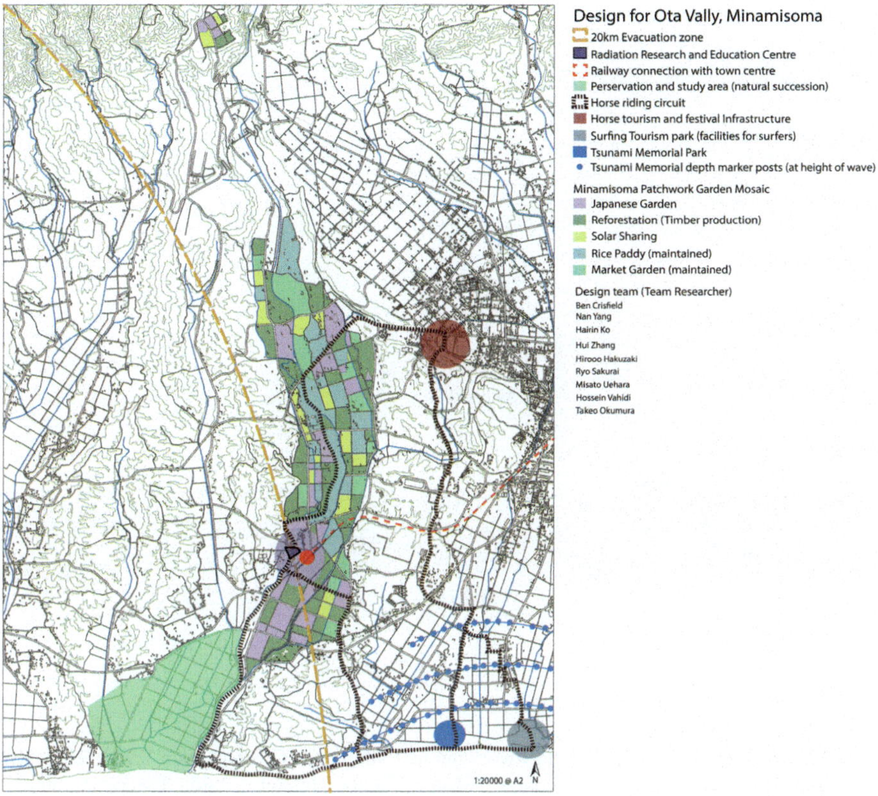

Fig. 6.11 Patchwork of different land-uses in Ota valley (Ben Crisfield, Nan Yang, Hairin Ko, Hui Zhang, Hiroco Hakuzaki, Ryo Sakuray, Misato Uehara, Hossein Vahidi, Takeo Okamura)

As an additional layer over this patchwork, several hubs are proposed, each with their own purpose (Fig. 6.12). The *Surfing Tourism Park* gives room to the already used beach site and encourages the people of Minamisoma to learn and enjoy surfing. This is an opportunity to bring young people to Minamisoma and encourage tourism and new tourism-related industries. It can also support the development of a new image to the town. The *Tsunami memorial shrine* is a larger built facility where both locals and visitors to the area can pay their respect. *Tsunami depth markers* of 5, 10 and 15 m above sea level visualise the scale of the disaster in the Ota Valley. Posts would be positioned 100 m apart and be a permanent reminder of the coastal lands for a tsunami. *Preservation areas* are created where high levels of radioactive pollution exist and natural succession can take place. These zones are located within the 20 km exclusion zone. These areas should be untouched and natural succession is allowed to occur without interruption. This would provide also a study site for scientists at the *Decontamination Museum and Research Centre* located in the area surrounding the old railway station.

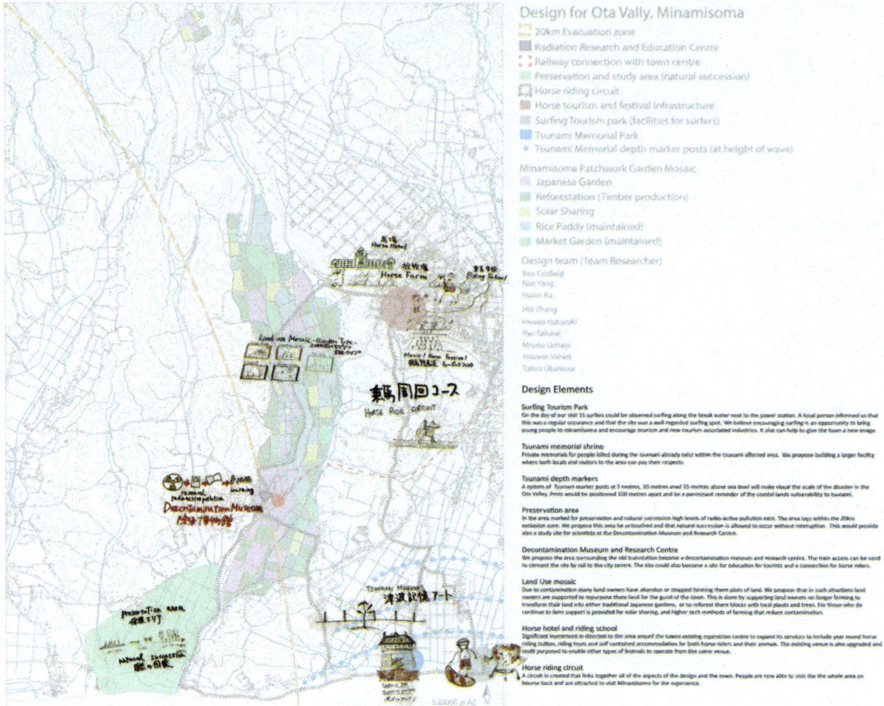

Fig. 6.12 Set of hubs on top of the patchwork in Ota Valley (Ben Crisfield, Nan Yang, Hairin Ko, Hui Zhang, Hiroco Hakuzaki, Ryo Sakuray, Misato Uehara, Hossein Vahidi, Takeo Okamura)

The train access can be used to connect the site by rail to the city centre. The site could also become a place for the education of tourists and a connection hub for horse riders.

The *Land Use Mosaic* emerges due to the contamination, which causes many landowners to abandon and stopped farming their plots of land. In this situation landowners should be supported to repurpose their land for the good of the town. This is done through supporting landowners to transform their land into either a traditional Japanese garden or to reforest these blocks with local plants and trees. For those who do continue to farm support is provided for solar sharing and higher tech methods of farming that reduce contamination. Significant investment is directed to a *Horse Hotel and Riding School*, located in the area around the towns existing equestrian centre, which can expand its services to include year round horse riding tuition, riding tours and self-contained accommodation for both horse riders and their animals. The existing venue is also upgraded and multi-purposed to enable other types of festivals to operate from the same venue. A *Horse Riding Circuit* is created to link all the hubs of the design and the town together. People are able to visit the entire area on horseback and are attracted to visit Minamisoma for the experience.

6.11 The Participatory Design Workshop

The residents' design workshop was held at the end of the design charrette process. The findings of the design team were presented in a presentation to the group, then a plasticine model was created to show the impacts of the design in three dimensions, and finally a participatory design workshop was held with citizens.

6.11.1 Presentation Preparation

In the preparation process for the presentation the overall design process was reviewed, which helped to clarify the message of the design proposal.

Through the 30-30 Exercise, the flow of the process became clear. Participants had the opportunity to '*Rethink*' modern society's vulnerability, which was exposed so clearly by the nuclear accident and the tsunami. The design for reconstruction from the disaster presents the opportunity for the regional natural and cultural 'genetics' to be '*Reborn*' After rethinking and rebirth, the '*Return*' can occur. By *return* two aspects are meant. First, the people can return to a way of living in a sustainable society, and second, people could physically return to their land. This conceptual approach was presented to the residents, and then, in order to incorporate the residents' perspectives, three-dimensional models were created using plasticine to visualise the outcomes.

6.11.2 Visualising the Design Concepts Using Three-Dimensional Models

Plasticine is useful as a tool to express the ideas of participants (Roggema et al. 2014). Using plasticine, which is available everywhere, participants could visualize the connections between the spatial components of identity in the region, and they could design their desired configurations of added and existing landscape elements. The landscape elements that were designed on paper suddenly become three-dimensional. Anyone can easily join, because of the common childhood experience of playing with clay, the material is inexpensive, and it facilitates dynamic communication.

6.11.3 Design Workshop

A participatory design workshop was held at the Ota Community Centre. The participants consisted of the same groups as in the 30 × 30 Exercise. Before the workshop, the design proposals, the landscape plan, and a demonstration of the plan by plasticine models were presented and exhibited at the venue. At the start of the workshop, the design team gave a presentation on the theme of '*Rethink, Reborn, Return*'. After questions and answers, participants were divided in four groups and asked to use the plasticine to express how they envision a plan for reconstruction, on top of aerial photos. Because of the fact that the participants were informed and saw the experts' designs and the three-dimensional models in advance, the workshop went smoothly. Students, workers, and seniors, all generations from various occupations, worked together as unified teams to think about the region's future, expressing their ideas by use of plasticine and placing the items physically on the aerial photograph. It was a dynamic exercise, with participants adjusting each other's creations and their spatial relationships. Some participants asserted their ideas strongly and sought agreement from everyone, while others yielded more and made an effort to consolidate the discussions. All of the teams were keen to develop their plans. This session was an important opportunity for communication and to understand what others were thinking.

At the end, each team leader made a presentation, explaining the spatial alignment and proposal developed by the group. For many students from local high school this was the first time to participate in such a design game and to speak in front of a group of people. They proudly introduced their team's work and they felt happy with such a positive experience.

6.12 Discussion and Conclusion

For this project, the design charrette method was introduced to develop a landscape design for the reconstruction of a region dealing with radioactive pollution. The process consisted of these steps:

1. Collection of local information
2. Identification of regional "genetics"
3. Landscape design by expert team
4. Preparation for presentations
5. Design workshop with local participation.

In this process there was close collaboration between residents and experts, and this made it also a mutual learning process. The time was brief, but there were several factors that made the workshop a success:

1. Information and knowledge sharing beforehand and during the first-day site visits
2. Communication and mutual understanding between experts and local participants through the process
3. Expert teams with diverse backgrounds and techniques

4. Easy communicative tools, including maps, aerial photographs, and plasticine models
5. Informative 30-30 Exercise and visualization in design proposal.

The design proposal, which the experts developed, aimed for a model of sustainable development rooted in the natural environment, the regional "genetics," the community, and economic resources. The participation of residents, while they were in the midst of reconstruction, was facilitated by maps and aerial photographs, and supported with three-dimensional models to express their diverse ideas of shrines, traditional festivals, local industries, biomass, solar power, collective operations, and more.

The participatory design workshop delivered informative learning effects, because it was held after the expert presentations and explanation of the three-dimensional models. While being in the midst of difficult challenges of the reconstruction process, all residents had a strong interest in the future of their region, and had a strong desire and commitment to participate. However, the constraints in knowledge and funding made it difficult for them to produce coordinated activities and results so far. This is where the assistance of experts and participation of a variety of people brings a special significance to the effort.

The workshop provided a platform for communication amongst the residents. During the workshop, participants could speak about shared ideas for the region and for the future. The result was the 30-30 map, on which all future ideas are summarized in a landscape plan with three-dimensional plasticine models. Experts using their knowledge did nothing more than help the participants to give form to their ideas, hopes, and desires. This process also enhanced communication amongst residents. After they saw their visions portrayed through this design, they said that their confidence and hopes had been restored. Therefore, this design charrette initiative served as a positive next step in building a resilient community.

The Charrette workshop was an effective tool to create a future vision, by bringing together domestic and international experts, non-profit organizations from inside and outside the region, government personnel, residents, and students. It was effective in harvesting observations about the current status of reconstruction from various angles, collecting opinions through talking about the future, and spatially visualizing the many ideas and approaches.

However, because the target region for this charrette is an area contaminated with nuclear radiation, there are uncertainties about the feasibility of plans. Soil contamination can flow, particularly from hilly areas, and the impacts on irrigated land are unclear. In May 2014, during a preparation visit for the workshop, *Asahi Shimbun* reported that the cesium levels from the rice grown in a trial paddy field in the Ota region in 2013 exceeded detection standards (Asahi Shimbun 2014). Media reported that this was an impact of particles dispersed from the treatment of rubble from the Fukushima Daiichi nuclear plant. The city and experts investigated but the actual facts were not confirmed. Again, regarding the future population, the net loss of population has gradually slowed, and some young people have returned to their hometowns. Whether or not they stay might require various approaches, including education, social services, and employment. These issues will require responses over time, integrated into future reconstruction initiatives.

The exciting design process generated very much support from the local population, as it was tactile, engaging, creative and visionary. Binding and co-creation is engaging people. Engaged people are sooner to contribute than the disengaged.

The embedding of the comprehensive plan in the philosophical tradition of Ninomiyasongtoku, allows people to be reborn, upon which rethinking about the land-use and functionalities of the landscape could take place. This helps to give back pride to the people and makes it attractive again for farmers, who could use their skills in the algae fields, vertical farms, and keeping horses.

The proposition are seen as realisable, are not too far from reachable proposals and close enough to current policies of Minamisoma prefecture.

The area could, when the proposals are implemented, become the most sustainable area of Japan. However, it may take 40 years to realise it.

References

Armstrong, R.C. 1912. *Just before the Dawn: The life and work of Ninomiya Sontoku*. New York: Macmillan.

Asahi, Shimbun. (2014). Yoshida Report, http://www.asahi.com/special/yoshida_report/. Last access date 24 Apr 2016 (in Japanese).

Batterbury, S. 2010. *Adaptive learning—A think-tank on preparedness for climate change adaptation in local and state planning in Victoria*. Melbourne: VCCCAR and University of Melbourne.

Bergdoll, B. 2011. *Rising currents. Projects for New York climate change adaptation in local museum of Modern Art*. New York: Museum of Modern Art.

Berkes, F., and C. Folke. 1992. A systems perspective on the interrelations between natural, human-made and cultural capital. *Ecological Economics* 5: 1–8. doi:10.1016/0921-8009(92)90017-M.

Campbell, J. 2013. Looking to the past, uncertain about the future: Japan's politics in an era of disaster and stagnation. *International Affairs* 89 (4): 1003–1017.

Carpenter, S.R., and L.H. Gunderson. 2001. Coping with collapse: Ecological and social dynamics in ecosystem management. *BioScience* 51: 451–457. 10.1641/0006-3568(2001) 051[0451:CWCEAS]2.0.CO;2.

Condon, P.M. 2008. *Design charrettes for sustainable communities*. Washington DC: Island Press.

Condon, P., D. Cavens, and N. Miller. 2009. *Urban planning tools for climate change mitigation*. Cambridge MA: Lincoln Institute of Land Policy.

Cyranoski, D., and G. Brumfiel. 2011. Fukushima impact is still hazy. *Nature* 477 (7363): 139–140. doi:10.1038/477139a.

Eagen, W.M., O. Ngwenyama, and F. Prescod. 2006. The design charrette in the classroom as a method for outcomes based action learning in IS design. *Proceedings of the Information Systems Education Conference, ISECON*, 23 (November). Available at: http://www.scopus.com/inward/record.url?eid=2-s2.0-84870457586&partnerID=40&md5=e5c2d8f17245ba7d211e5ef2d3e08943

Endo, S., T. Kajimoto, and K. Shizuma. 2013. Paddy-field contamination with 134Cs and 137Cs due to Fukushima Dai-ichi Nuclear Power Plant accident and soil-to-rice transfer coefficients. *Journal of Environmental Radioactivity* 116: 59–64.

Folke, C., J. Colding, and J. Berkes. 2003. Synthesis: Building resilience and adaptive capacity in social-ecological systems. In *Navigating social-ecological systems: Building resilience for complexity and change*, ed. F. Berkes, J. Colding, and C. Folke, 352–387. Cambridge, UK: Cambridge University Press.

Folke, C., T. Hahn, P. Olsson, and J. Norberg. 2005. Adaptive governance of social ecological systems. *Annual Review of Environmental Resources* 30: 441–473.

Furutani, T., K. Uehara, and J. Murai. 2012. A study on community-based reconstruction from nuclear power plant disaster-A case study of Minamisoma Ota Area in Fukushima. *Journal Ref: Journal of Disaster Research*, 7, no. 1, 1–2. Retrieved from http://www.fujipress.jp/finder/xslt.php?mode=present&inputfile=DSSTR000700070003.xml

Girling, C., R. Kellett, and S. Johnstone. 2006. Informing design charrettes: Tools for participation in neighbourhood-scale planning. *The Integrated Assessment Journal*, 6, no. 4: 109–130. Available at: journals.sfu.ca/int_assess/index.php/iaj/article/download/238/218

Glenn, J.C. 1994. *Participatory methods*. Washington DC: United Nations University.

Goedert, J. 2008. Rebuilding new orleans With Affordable, Residential Construction. *Cityscape: A Journal of Policy Development and Research* 10 (3): 167–184.

Hirano, K. 2013. Difficulties in post-tsunami reconstruction plan following Japan's 3.11 Mega disaster: Dilemma between protection and sustainability. *Journal of Japan Society of Civil Engineers, Special Topic—2011 Great East Japan Earthquake* 1: 1–11.

Howard, J., and K. Hurst. 2009. *Planning for climate change mitigation and adaptation in North Central texas-a roundtable discussion*. Arlington: School of Urban and Public Affairs, University of Texas.

Ishikawa, M. 2013. Developing Resilient Infrastructure as a basis for restoration from the Great East Japan Earthquake. *International Perspectives in Geography* 1: 165–208.

Ishikawa, M. 2014. New trend of landscape planning and design in Japan. *Urban and Landscape Perspectives* 17: 27–56.

Kabashima, H. 2012. Current Issues in legal policy for recovery from the aftermath: One year after the 3.11 Tôhoku Earthquake and Tsunami. In: *Socio-Legal Norms in Preventing and Managing Disasters in Japan: Asia-Pacific and Interdisciplinary Perspectives, March 1–2, 2012, Sydney*. Available at: http://sydney.edu.au/law/anjel/documents/2013/ZJapanR_34_06_Kabashima.pdf

Luz, F. (2000). Participatory approaches in landscape ecology—a basis for acceptance and implementation of concepts for managing multifunctional landscapes. In Tress *Conference material for the conference on "multifunctional landscapes", Centre for Landscape Research, Roskilde*, ed. J. Brandt, B. Tress, and G. Tress, 264. Denmark: Centre for Landscape Research, University of Roskilde.

Nakano, M., and P.P. Povinec. 2012. Long-term simulations of the 137Cs dispersion from the Fukushima accident in the world ocean. *Journal of Environmental Radioactivity* 111: 109–115. doi:10.1016/j.jenvrad.2011.12.001.

Okada, A., and S. Kurosu. 1998. Succession and the death of the household head in early modern Japan: a case study of a Northeastern village, 1720–1870. *Continuity and Change* 13 (1): 143–166.

Roggema, R. 2009. *DESIGN, Final Report Hotspot Climate Proof Groningen*; Groningen/Wageningen/Amsterdam: Province of Groningen and Climate Changes Spatial Planning, 159 pp.

Roggema, R. 2013. *The design charrette: Ways to Envision sustainable futures*. Dordrecht, Heidelberg, London: Springer.

Roggema, R. 2015. Reborn, Rethink, Return to Minamisoma, Japan. *Adjacent Government* 5 (2015): 22–25.

Roggema, R., L. Vos, and J. Martin. 2014. Resourcing local communities for climate adaptive designs in Victoria, Australia. *Chinese Journal of Population Resources and Environment*, 12, no. 3: 210–226. Available at: http://dx.doi.org/10.1080/10042857.2014.934951.

Rossi, V., E. Van Sebille, A. Sen Gupta, V. Garçon, and M.H. England. 2013. Multi-decadal projections of surface and interior pathways of the Fukushima Cesium-137 radioactive plume.

Deep Sea Research Part I: Oceanographic Research Papers 80: 37–46. doi:10.1016/j.dsr.2013.05.015.

Sakai, M., T. Gomi, M. Nonokawa, T. Wakahara, and Y. Onda. 2014. Soil removal as a decontamination practice and radiocesium accumulation in tadpoles in rice paddies at Fukushima. *Environmental Pollution*, 187, 112–115. Available at: http://linkinghub.elsevier.com/retrieve/pii/S0269749114000086

Samuels, R.J. 2013. Japan's rhetoric of crisis: Prospects for change after 3.11. *The Journal of Japanese Studies*, 39, no. 1: 97–120. Available at: http://muse.jhu.edu/content/crossref/journals/journal_of_japanese_studies/v039/39.1.samuels.html. Accessed 2 Apr 2014.

Shineha, R. and M. Tanaka. 2014. Mind the Gap: 3.11 and the Information Vulnerable. *The Asia-Pacific Journal: Japan Focus*, 12, no. 7. Available at: http://japanfocus.org/-Tanaka-Mikihito/4076/article.pdf.

Solnit, R. 2010. *A Paradise Built in Hell: The Extraordinary Communities That Arise in Disaster*, Penguin. Available at: http://www.amazon.co.jp/Paradise-Built-Hell-Extraordinary-Communities-ebook/dp/B003XQEVLM/ref=sr_1_4?ie=UTF8&qid=1447652705&sr=8-4&keywords=rebecca+solnit. Accessed 16 Nov 2015.

Suzuki, K. 2013. important cultural landscape list and the decay o f traditional agricultural life: A case of dispersed settlement in the tonami Plain, Japan. In: *Globalization and new challenges of agricultural a nd rural systems : Proceeding of the 21st Colloquiu m of the Commission on the Sustainability of Rura l Systems of the International Geographical Union (IGU): Nagoya,* 59–70, *August 2013*.

Suzuki, I. and Y. Kaneko. 2012. Brief Overview of the Tohoku 3.11 Mega Earthquake, Tsunami, and Fukushima NPS Disasters. *Public Administration, Governance and Globalization*, 4, 7–23. Available at: http://www.springerlink.com/index/10.1007/978-1-4614-6151-7. Accessed 31 Dec 2015.

Tanaka, T., D.B. Abramson, and Y. Yamazaki. 2009. Using GIS in community design charrettes: Lessons from a Japan-U.S. collaboration in earthquake recovery and mitigation planning for Kobe. *Habitat International*, 33, no. 4: 310–318. Available at: http://dx.doi.org/10.1016/j.habitatint.2008.08.006.

Valencia-Sandoval, C., D.N. Flanders, and R. Kozak. 2010. Participatory landscape planning and sustainable community development: Methodological observations from a case study in rural Mexico. *Landscape and Urban Planning* 94: 63–70.

Walker, J.B., and M.W. Seymour. 2008. Utilizing the design charrette for teaching sustainability. *International Journal of Sustainability in Higher Education* 9 (2): 157–169.

Wilkinson, C. 2011. Social-ecological resilience: Insights and issues for planning theory. *Planning Theory* 11 (2): 148–169.

Zhang, H., W. Yan, A. Oba, and W. Zhang. 2014. Radiation-Driven Migration: The Case of Minamisoma City, Fukushima, Japan, after the Fukushima Nuclear Accident. *International journal of environmental research and public health*, 11, no. 9, 9286–9305. Available at: http://www.ncbi.nlm.nih.gov/pubmed/25207491. Accessed 13 Sept 2014.

Chapter 7
Planning and Design in Kesennuma: Remember, Reconnect, Reform

Wanglin Yan, Rob Roggema and Luke Middleton

7.1 Background

Before the 2011 disaster, Kesennuma was a key city on the Sanriku coast, with a population of 75,000. Major industries included fishing, marine products processing, and shipbuilding. While the ria geography of the coast provides a wealth of natural resources, disaster risk comes along with it.

The Great East Japan Earthquake of 2011 was a multiple disaster that occurred in the midst of a complex domestic and international situation. In this disaster, the earthquake registered 6 on Japanese seismic scale in the central part of Kesennuma, and 5 in the Motoyoshi area in the south part of the city. Inundation from the tsunami reached a depth of 19.3 m in the Motoyoshi District, it swept over the Naiwan, Shishiori, and Minami-Kesennuma districts and carried large boats up onto the land. Damaged oil tanks exploded and caused major fires in the Shishiori village. Losses were significant for the city, with 1359 dead and 200 missing, plus 108 subsequent disaster-related deaths, 15,815 houses destroyed, and 9500 families affected.

Half a year after the disaster, the city created the Kesennuma Disaster Reconstruction Council, consisting of members of the urban planning council and experts who came from Kesennuma, and undertook efforts to form a reconstruction

W. Yan (✉)
Faculty of Environment and Information Studies, Keio University,
Fujisawa, Kanagawa, Japan
e-mail: yan@sfc.keio.ac.jp

R. Roggema
Faculty of Design, Architecture and Building, University of Technology Sydney,
Ultimo, NSW, Australia
e-mail: rob.roggema@uts.edu.au

L. Middleton
EME Design, Melbourne, Australia
e-mail: luke@emedesign.com

R. Roggema and W. Yan (eds.), *Tsunami and Fukushima Disaster: Design for Reconstruction*, DOI 10.1007/978-3-319-56742-6_7

plan based on expert input. At the same time a residents' committee for reconstruction was established to listen broadly to residents' input and incorporate that input into the reconstruction plan. The recommendations of both committees were summarized and the Kesennuma Ten-Year Reconstruction Plan was adopted.

The city and its port have since been recovering slowly (Japan Guide 2015). The rubbish was cleaned, and the road and supply infrastructure is reconstructed. Most of the residents with detached family houses have rebuilt their houses at nearby higher lands or are listed for public condominiums. The low lands in city centre were readjusted and elevated mainly for industrial use, parks and commercial activities.

In Kesennuma City, reconstruction of the 'hardware', such as city infrastructure, housing construction, and industrial facilities, is proceeding according to plan, albeit with some delays. Many of these projects are government-directed and project based. However, the purpose for each one is different and the department in charge is also different. The City hall is overwhelmed with the work of developing budgets for reconstruction funds and the implementation of projects, and does not have the time to think carefully about what the city will become as a whole. Neighbourhood councils were wiped out by the disaster, or the residents are not organized, so in many places there is no coordinated activity for community planning. The reconstruction outside of city centre is delayed because of limited resources and capacity on one hand, and the lack of consensus on the reconstruction among citizens and the municipality on the other hand.

For reconstruction it is necessary to provide urgent assistance so that survivors can quickly free themselves from the conditions of the disaster, and also to balance those needs with a future plan that considers sustainability. As approaches for that, we often see government-directed citizen-participation models, but in the race against time, reconstruction often does not advance the way people may have expected. Meanwhile, with the reconstruction from the 2011 disaster, many experts went to the local area and cooperating on reconstruction with diverse stakeholders, including universities, governments, citizens, and companies. In such a context, the government, citizens, and experts, thought together about the future, without distinction or hierarchy, and from that situation was born the potential to plans and design. This chapter refers to that approach as 'co-creative reconstruction and community planning'.

After the disaster, over 100 companies and at the peak more than 40 university research offices focused on Kesennuma and offered assistance with reconstruction. During reconstruction, there was no extra capacity to respond to all the offers, and many ideas proposed were left untouched. Also, in the confusion after the disaster, in many cases, the offers for assistance did not match the local needs.

As the situation settled down, more people were able to reflect and ask more deeply about the desired direction for reconstruction. The concentration of human resources and information was a tremendous resource, and there were now attempts to find ways to make use of them. The Association to Create a Livable Kesennuma City (*Kesennuma-shi Sumiyosa Sozo Kaigi*) was established for that purpose in 2013, with participation from citizens, experts, local businesses, and outside corporations. It set out to review the reconstruction proposals. Over 150 proposals were listed in categories such as social infrastructure, energy, industrial revitalization, and social welfare, and the best projects were selected. Companies were interviewed, the details of assistance and requests confirmed, and an effort was made to match decisions with reconstruction needs. In the end nothing turned into an actual project, but the effort was an opportunity to share information from government, citizens, and businesses from local and elsewhere, and to promote communication. It also confirmed that there was a need for a platform to act as intermediary for assistance from outside the region.

Similar moves were also evident in other disaster-affected municipalities. In Higashi-Matsushima, an intermediate assistance organization named HOPE was established, and it is promoting projects involving citizen and corporate participation. In Kesennuma as well, after extensive discussion, it was decided to establish the Organization for a Liveable Kesennuma City (*Kesennuma-shi Sumiyosa Suishin Kiko*, referred to below as the 'Sumiyosa Organization'). It was established in 2014, and developed projects in three major areas: energy, healthy society, and industrial revitalization. Examples of projects include:

(1) Matsuiwa district reconstruction and community planning project.
(2) Project to introduce Manson Energy Management Systems (MEMS) for emergency public housing.
(3) Marketing channel development project for local products.
(4) U-turn and I-turn project for human resources development.

The objectives of the four projects, respectively, are reconstruction planning based on citizen participation, using smart technologies for emergency public housing, assisting local small and medium enterprises, and training the younger generation, who people hope will settle in the area. The projects have just recently begun. They are aiming to share information internally and externally, and to facilitate assistance from governments and corporations. The Matsuiwa district reconstruction and community planning activities arose from the activities of this organization.

The District of Matsuiwa is the target area of the design charrette and is such an area. The Matsuiwa district is located at five kilometres south of Kesennuma City centre on the north side of the Omose River. The Omose District is on the south side. In the 2011 disaster, considerable damage from the tsunami was focused in the Katahama area of Matsuiwa District and the Ozaki area of the Omose District. The tsunami washed the houses along the coast and the river away. The Shrine of Ozaki at the cape survived and saved more than 30 evacuees. Enunkan, a historical house owned by the Ayukai family since the Edo era is currently the home of Mr. Naofumi Ochiai, and is famous in Japanese literature in the Meiji era. This house is located at the hill in front of the bay, with a beautiful Japanese garden and a nice view of the harbour. This specific location made it possible to witness the tsunami in all its strength. The impact of the tsunami is best visualised when comparing the

Fig. 7.1 The areal view of Kesennuma before and after the tsunami

areal views from before and after the tsunami (Fig. 7.1). All houses in the entire coastal zone were diminished.

In a public opinion survey conducted here immediately after the disaster, 40% of respondent households wanted to resettle outside their district, 20% wanted emergency public housing, 30% were still considering on returning, and 10% wanted to rebuild locally. Because only a small proportion wanted to rebuild locally, the city designated approximately 44.5 hectares at the mouth of the Omose River as ‘disaster risk area’. With the Omose River as a dividing line, the north side was designated for use as a low-lying zone, and the south side as a disaster prevention green area. Seawalls and dikes were built in the coastal area and Omose River, with the plan to prevent inundation damage from a level one tsunami. For dikes on the Omose River, the prefecture was in charge of the project, and reconstruction was done to a point 600 meters upstream from the river mouth. The seawall was approximately two meters above sea level, but the plan is to raise this to between 4.0 and 7.2 m with a foundation 20 to 30 m wide (Nikkan Kensetsu Shimbun 2012).

With the desire to live here being so low, there was no need for a big rush for reconstruction. After four years, there was no organised activity for town planning in this area. Though no decisions had been made on how this land would be used, restoration work on the fishing port and prefectural road was under way. It was decided to build a sports park in the Ozaki area of the Omose District, and beside that along the Omose River a seawall 7.2 m in height is under construction. Only the civil engineering works were moving ahead, though there was no coordination between projects and districts. Some surrounding residents who were concerned about the status of reconstruction in the area started approaching authorities. The Sumiyosa Organization incorporated the citizen’s input and created projects, starting by gathering residents and experts together from near and far, and holding a design workshop.

7.2 Aims of the Design Charrette

With such a good location and rich cultural-natural resources, the residents, the landowners and citizens are interested in rebuilding the district as a zone of recreation, learning, and communication. This might be the only chance for the city to reconstruct the area with full participation of the public in a bottom up process. Citizens have started to talk and have developed some ideas with each other. It is unknown how to represent the ideas professionally in harmony with the landscape and culture. It is also unknown how the ideas could be implemented and to contribute to the future of the city. The aims of the design charrette were as follows:

(1) To understand the demands, wishes and interests of the people (residents and outsiders);
(2) To work together with the people in co-creation (residents and outsiders);
(3) To develop a future vision for the region;
(4) To design the pathways and spatial alignments of Matsuiwa District;
(5) To make recommendations for implementation of the plan and the design.

From 31 August 31 to 3 September 2015, a design charette, entitled 'Thinking about the future of Matsuiwa and Omose Districts', was organized by a team of educators and students from Keio University, Swinburne University of Technology, and EME Design. People arrived on August 30, and on August 31 they visited the area and had interviews during the day. That evening, 20 local residents participated in what is known as a '30-30' exercise, which was the first opportunity since the 2011 disaster for residents of Matsuiwa and Omose Districts to get together for a broad discussion about reconstruction plans. During the first half of the workshop, participants were divided into four groups to recall what the area was like thirty years ago in terms of economy, population, culture, and climate. During the second half of the evening, the groups developed an image of their town thirty years in the future. The results of the '30-30' exercise are summarized in Table 7.1. A number of features can be observed in this Table.

1. The residents are proud of the natural and cultural resources in their area. They want these natural and cultural resources to be a part of reconstruction.
2. They see both sides of the Omose River in a unified way, irrespective of administrative jurisdiction.
3. The residents are not only interested in the fishing port and river mouth areas, but also in areas further upstream along the Omose River.
4. The region would like to see significant development happen, because transportation access is expected to improve with improvements to the Sanriku roads and prefectural roads.

Through this workshop, the residents were able to discover their shared attitudes regarding reconstruction, on topics they had until then been considering alone or discussing only in a fragmented way.

Table 7.1 Results of '30-30' exercise (prepared by Takahiro Kanamori)

30 years ago	30 years in future
Society/community	
Hachiban Shrine Matsuri (festival)	Lovers' walk
Toko Matsuri (festival), Bon-odori (summer event with traditional dancing)	Regional museum
Enunkan (historical house)	Village of folklore (Nekobuchi, courtesans)
Clamming (collecting shellfish)	Road of stories
Daimyo procession at Ozaki Jinja Shrine	Bunkyo district
Many children lived here	Toko Matsuri (festival)
Economy/agriculture/industry	
Kesennuma/Katahama supermarket	Shopping district, commercial facilities
Traditional entertainment district	Hotels, "minshuku" (inns)
Aquaculture industry	Companies, seafood processing plants
Combination livelihoods (farming, fishing)	Cafes
There was no bypass road	
Many rice paddies	
Climate/nature/disasters	
Chile tsunami	Nature learning (aquaculture, etc.)
Flooding from high tides	Cycling routes
Every family had a "danbeibune" loading barge	Tourist boats, yacht harbor
Could skate on ice	Sports park, multipurpose grounds
Flooding and sewer overflow from heavy rains	Beaches (recreational)
Summers were cool	Cherry trees

7.3 Criteria for the Design

The landscape design process for reconstruction involved taking available government and historical information about the local area, input from residents and various reconstruction projects, and arranging it all conceptually and spatially. On 3 and 4 September, the design team worked to create a design proposal. Consideration was also given to local characteristics and future objectives. Project criteria were established in the reconstruction plan based on site conditions of the Matsuiwa and Omose Districts, history and culture, and residents' requests.

7.3.1 *Disaster Resilience*

The Matsuiwa side is in a low-lying zone, and was severely damaged by the tsunami. It is also frequently flooded by rainwater from inland areas. Before the disaster, inland flooding was addressed with ditches, pumping stations, and floodgates. There are also local voluntary disaster prevention organisations, which pay attention to disaster countermeasures. In reconstruction, a priority should be placed on safety.

7.3.2 *Urban Infrastructure*

Infrastructure construction is advancing in a broad area including seawalls, prefectural roads, Sanriku roads, and bus rapid transit (BRT), and for these the foundations are being built higher. However, infrastructure improvements lack the perspective of looking at it as a town. There is a need to create an overall vision that takes advantage of local characteristics.

7.3.3 *Industrial Revitalization*

Before the disaster, industrial districts were concentrated on the Omose side of the district, and the Matsuiwas side was mainly housing. The city is envisioning industrial land in the reconstruction plan, but has not developed a concrete strategy to attract industry. It is necessary to promote a strategy to attract business from the perspective of excellent siting conditions.

7.3.4 *Environmental Harmony*

Residents have a strong desire for harmony between the natural environment and reconstruction projects. The district has cultural/historical sites, such as Enunkan, Hachiban Jinja (shrine), and Ozaki Jinja (shrine), natural landscapes of Oshima Island, ecosystems of the fishing port, river mouth and upstream areas of the Omose River, and industries, such as the fishing industry along the coast, agriculture along the river, and forestry further inland. It is special to have such diversity close to the centre of Kesennuma City.

7.3.5 *Health and Welfare*

Survivors from the disaster have been relocated outside the district, but still live in the surrounding area. A group resettlement site is being constructed along the Omose River, three kilometres upstream. Improvements are needed in the living environment, with due consideration for the health and welfare of these people. Reconstruction projects should consider how to perform community planning in a way that it creates health and liveability for the people being resettled due to the disaster, and people who already live there.

7.3.6 *Education and Learning*

The Omose Elementary School and Omose Middle School are in the area. They have been enthusiastic about nature and local community learning programs, and have been recognised as innovative examples of United Nations initiatives in education for sustainable development (ESD). The Enunkan historical house is being used by the Agency for Cultural Affairs as a training site for personnel to learn about Japanese traditional gardens. The Ozaki Jinja shrine is the location of a memorial for 30 people who survived the tsunami. In the hilly areas there are projects to encourage the use of the multiple functions of forests, and this provides good material to learn about ecosystem services. The fishing port provides opportunities to observe and study seafood processing. There is significant potential to expand each of these resources into programs for nature education, disaster prevention education, industrial education, and local community education.

7.3.7 *Regional Collaboration*

Villages in the district are mutually connected through kinship, regional bonds, and beliefs. The disaster has separated many of the survivors, but they continue to have a strong interest in their hometown. Ideally, their involvement in reconstruction should not be just as the original residents, but as citizens of Kesennuma City, and reconstruction should be promoted as a collaborative creation of new and original residents.

7.4 Detailed Description of the Design

This area has a rich natural landscape and historical culture, and it has the potential to become a new town through reconstruction.

The Enunkan house on the hill represents a cultural story from the past. Literally En-Un-Kan means the house of smoke and clouds. Nature is seen as a counterpart of man, as a contained autonomous force with shaping power. It reflects the respect of man for ecology, air, water and earth. From every miracle the smoke goes up in the clouds (water from the river evaporates into the clouds), the clouds move towards the mountains and when it rains in the mountains, the yellow river brings the miracles back. The river is yellow because Sen (God) brings you value from the gold mine, up in the hills. This colours the river yellow and brings the water from the mountain to the sea. This refers to the Yellow River in China where the Emperor has 1000 beautiful girls. So in Enunkan you can imagine you are the Emperor and the girls will appear, using the water of the river as perfume. Halfway the river there is a flat stone, which the girls use as a mirror before she can show up at the Emperors house. Kesennuma refers also to this story. *Ke* means miracle smoke, *sen* means God or knowledgeable and *numa* means wetland or water. The water is for Kesennuma as the perfume for the beautiful girls. Therefore the river is called Oinan, which is the same as the word for beautiful girl.

In the hinterland there are much more stories related to this. There is a cat, hidden in the woods and represented by a special rock, eating the beautiful girls. There is a little shrine for a god of safety, helping you (the beautiful girls?) to make love without any diseases. Additionally there are shoes-like stones in the river (lost by the girls?) and a smile face under a bridge (to keep her happy?).

So, there is a connection between house of the ‘emperor’ at the coast and the mountain, also representing a cycle of smoke, clouds, rain, water, river, wetland and sea.

This story has been the basis of the plan for Kesennuma. The relation between the sea and the mountains is strengthened through new connections for bikes and making abundant spaces for the river to show its natural habitus. The foundation of the plan is to intensify the exchange between mountain and water, both through interfering recreational flows and the conjunction of sea and land as a dissipative structure, allowing water deep in the hinterland (Fig. 7.2). This way the history will be *Remembered.* Not only the cultural history, but also recent history of the tsunami, through *Reconnecting* the people with these historic and recent stories and reconnecting the mountainous area with the sea. In order to make tis manifest the area will *Reform.* Not only the area hit by the tsunami, but also deep in the hinterland.

The tsunami deserves a new layer in history, where the disaster can be remembered and a future lookout can be generated. Again, the relationship between the river and the sea can be made manifest.

The area is loaded with connections. Between the sea and the mountains, between the fish in the water (Kesennuma is an important fishing port) and the cat. Between the Emperor and the beautiful girl, between the water as an enemy and a friend. Between the old house and the shrines, between historic stories and future developments. All these connections reappear in the design for the area.

Fig. 7.2 The regional plan for Kesennuma: Remember, Reconnect, Reform

7.4.1 The Coast

This region is vulnerable to both tsunamis and floods. Seawalls protect the land from the sea, pumps remove floodwaters, and river dikes control the flow of the river—but for all of these, water is seen as something unwelcome. In the design exercise there was the idea to reverse this thinking, stop rejecting, and instead look at how to *accept* water. The low-lying zone could be utilized as-is as a zero-meter elevation area, or even dug out slightly to create a flood control basin. On the riverbanks, instead of using concrete dikes, the watercourse could be expanded using natural methods. Thus, even if a tsunami came, or if flooding occurred, the flood control basin could act as a buffer and damage could be mitigated. The area that had been dug out to create a water surface could be a water landscape, and during regular times be used as a safe place to enjoy water. The soil from excavating the flood control basin would not be transported far away, but could be used to build up both banks, and extend as a natural dike from the Hachiban Jinja shrine. On the dikes, trees could be planted and buildings built to create scenes that blend with nature.

The redesign of the coastal zone creates a larger dissipative zone, which is much better capable of attenuating wave power. The coastline is, as it were, perforated to allow the water from the sea inundate the hinterland. At the same time new mountainous areas are created near the coast, extending the mountain area to create safe havens. The result of tis system is a varied coastline of entering sweater and hilly extensions. In contrast with implementing one huge seawall this design

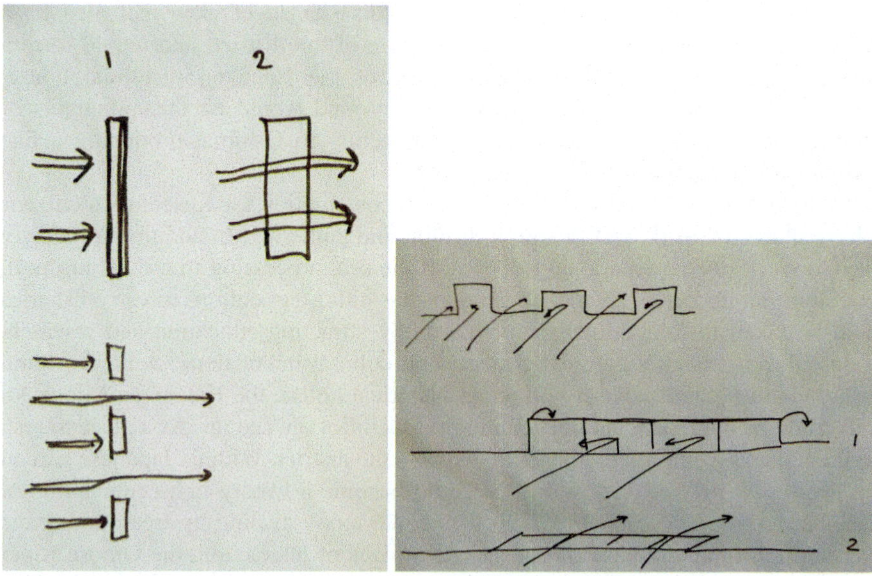

Fig. 7.3 Difference between a sea wall (1-2) and a dissipative structure

proposition provides a safer situation and creates an attractive coastline for tourism and wellbeing of the residents of the area. A seawall can be torn down by the power of the sea and the hinterland will flood, while a perforated seawall allows bits of the waves move through and holding back the rest of the water (Fig. 7.3). The structure will not be torn down and quiet water adds to the quality of life behind the wall. The water is seen as a friend, as it receives and downsizes the flood. Therefore water is given space in the plan.

7.4.2 History in the Making

Matsuiwa and Omose districts have a long history in Kesennuma City. In the past, the area was lively, with a gold mine in the mountains and hot springs nearby. The Omose River has clear water, and there are legends that traditional courtesans (known as Oiran) of times past would go to a mirror rock by the river to prepare their makeup. These kinds of stories are still told to this day, and provide content for learning about the community. Such stories never age, so they should be incorporated into reconstruction and community planning. A row of cherry trees could be planted along the tsunami and inundation line, leaving traces of the disaster, and a memorial stone could be placed on the hill of the Ozeki Jinja shrine, for this to be used as a place for disaster education. Reminiscent of the date of the disaster (March 11, referred to in Japan as "3.11") a public space could be created with dimensions 93.1 by 93.1 m^2, and a pre-disaster street view recreated on a one-tenth scale. The square would be the centre of the district, and would form a diamond shape with the three scenic spots of the Enunkan historical house, Hachiban Jinja, and Ozeki Jinja. In the background would be the mountains of Oshima Island, and the planned bridge connecting to Oshima, producing a fine landscape composition.

Multiple functions could be created after reconstruction for Kesennuma citizens to have places of work, rest, learning, health, and gathering. In addition to being a place to work (the conventional fisheries in the sea, processing in the fishing port, agriculture along the river, and forestry in the hills), agriculture-forestry-fisheries could be taken to four dimensions by actively attracting education and research, product development, nature experience, and cultural interactions. A new cultural facility could be built near the Enunkan historical house, the Katahama Minshuku (inn) could be upgraded, and the fishing port facilities opened up. As a place to rest, the Enunkan as a cultural facility is an icon for the district. With its Japanese garden and lovely row of trees below it, this could become a twenty-first century line of trees that represents Kesennuma. Cyclists could enjoy cycling upstream along the river through rural and pastoral scenes. As a place of interaction, the Omose River could become an important transportation route for residents who have resettled elsewhere, outside the district. On the route along the river, they could commute to work, and walk back toward the mountains. Residents of new housing developments could walk along the dikes toward the sea, get exercise and use their hands by gardening on the former sites of their homes that were damaged by the tsunami.

In order to make the tsunami manifest and make sure it will be remembered, several spatial ingredients are key parts in the design.

1. Cherry line (Fig. 7.4): The tsunami caused a flood, which reached far inland. A line of cherry trees will mark the line in the landscape to where the flood entered, and will be accompanied by a bicycle lane. This way people will always remember and are able to follow the line and this makes it possible to become aware how large the influence of the tsunami was on the landscape.
2. Monument: In the heart of the area exactly in the middle of three historic points in the landscape a monument for the lost settlement will be raised (Fig. 7.5). This monument consists of a map scale 1:10 of the former settlement. The size of the square is 93.3 × 93.3 m. The map will be constructed using debri materials to shape the old street pattern and houses. This way people can tae a walk through their old village, commemorating the place where they lived, but also the place where family of relatives and friends died. The monument is one of four anchor points in this area. Together with the Enunkan house, the temple up the hill and the shrine on the cape it form the anchor points of a pyramid, visually connected to each other because they're all located at elevated points in the landscape.

Fig. 7.4 The cherry line, marking the extent to where the flood came

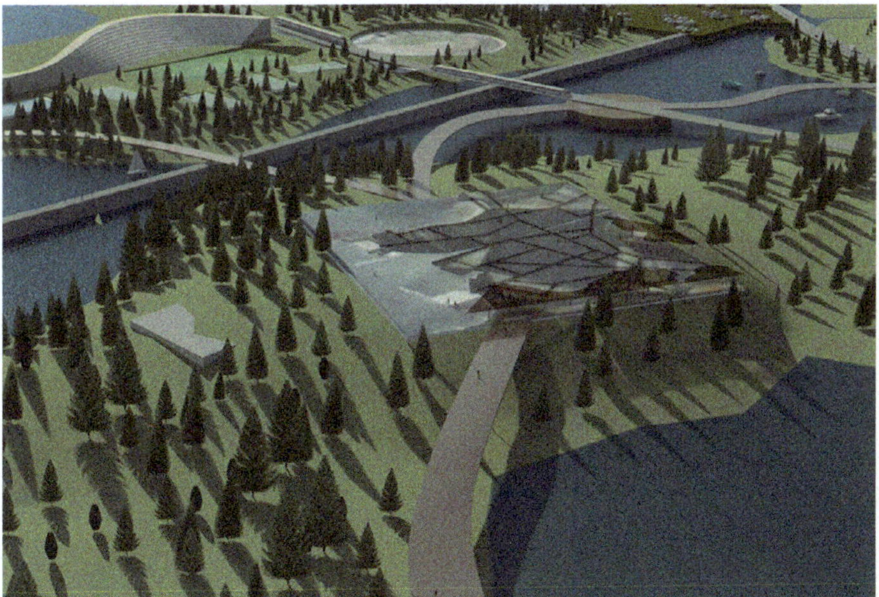

Fig. 7.5 The monumental square (EME design)

3. Excavated houses: On the edges of the central site the foundations of the houses are still visible. The proposal is to excavate these foundations and to make the remnants of the tsunami visible. An underground viewpoint will be added, from where the foundations, including soil, sewage and utilities can be viewed.
4. Infrastructure: the area will be well connected to its surroundings. There will be a new bus stop on the old railway track, the new straight levee, which this under construction is transformed to be used as a main access promenade and there will be multiple parking and walkways.
5. The beach side is the ideal location for a sports park (Fig. 7.6). Embedded in a hill, sports such as tennis, baseball, football and swimming find their spot in a forest. The amphitheatre is suitable for spectators of the sports, but also a southern oriented stand towards the beach. The beach itself is renewed and safeguarded behind a perforated former dam. This creates a tranquil and well-protected sea.
6. The hinterland is stronger connected to the seaside. The flooding from the mountains, in periods of heavy rain, is mitigated through widening the streams. Additional bike-paths, walkways and streams will strengthen the connectivity. Along the streams and pathways existing and new shrines are located, such as the cat-rock and the mirror stone. The people that have been relocated to higher grounds and lived near the coast before will be asked to start gardening and maintaining (the foundations of) their former house.

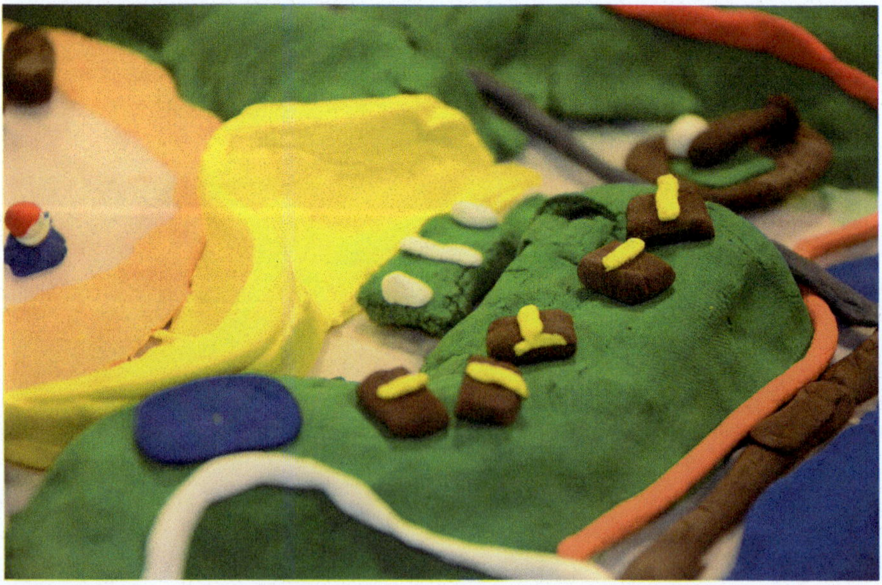

Fig. 7.6 Plasticine image of the sports facilities on the hill and the amphitheatre

Fig. 7.7 View of the Master plan (Luke Middleton)

The Master plan (Fig. 7.7) is a guide for the long term. It gives directions into the future, but doesn't describe in detail how every part should be designed and developed. Therefore, as a next step further detailing of designs is necessary, however fitting within the framework of the Master plan.

7.5 Residents' Design Workshop

In order to make designs easier to understand, a three-dimensional model was created using plasticine. Residents were brought together again on September 3 to resume the workshop. The design team presented and discussed the design proposal. In the second half, the reconstruction design was done with the residents. They were divided into four groups and made designs on their own (Fig. 7.8). By taking their ideas and expectations and put them onto a map, they could see how their ideas related to each other, then started to discuss and make trade-offs about the future direction for the town as a whole, and to adjust their ideas. Thus, they experienced themselves that town planning and landscape design are not only for experts, and that residents themselves are entirely able to do this.

They could sense that it was not necessary to have fixed ideas about reconstruction, and if they could dream bold ideas, various things might become possible. Someone said: 'At the first workshop, we were allowed to say anything we wanted, but I was not sure what would come of this'. Participants had some uncertainty and curiosity, but many people participated in the second workshop. They were surprised to see what had been fragmented information was now set down neatly in a spatial design. The residents had until this point only seen planning drawings and

Fig. 7.8 Kesennuma City, design workshop with resident participation (*Photo* Wanglin Yan)

construction drawings from government offices, but in this workshop, they encountered various ideas that they had not seen before, such as the bold concept of accepting water, the method of skilfully locating various ideas and desires in a spatial way, the systematic approach of looking at the river mouth and upstream in a unified way, and the enthusiasm that reconstruction was the start of a new historical story. 'If we can make this happen, surely our children will be happy'. 'I would be satisfied if we could realise even one-tenth of this'. Many comments from participants articulated experiences like this.

7.6 Conclusion

The design proposition for Kesennuma reflects the history of the place and connects this with anticipation of the future. The reconnection between the natural systems of the hills and the seaside is essential in creating a stronger resilience. The place will be used to remember the victims and the disaster of the tsunami through the monumental square and the excavated houses yet at the same time it embraces the potential impacts through welcoming the sea deep in the hinterland. This makes the area less vulnerable for eventual disasters and creates at the same time a very functional area, in which recreational activities dominate. Without the need of a high and disturbing seawall the area stays safe.

References

Japan-Guide. 2015. *Kessenuma*. http://www.japan-guide.com/e/e5036.html. Accessed 20 Jul 2015.

Nikkan Kensetsu Shimbun. 2012. 松岩・面瀬地区の市街地ほぼ全域を非居住へ(気仙沼市). http://www.jcpress.co.jp/wp01/?p=6900. Published 12 Apr 2012.

Chapter 8
Visualisation of Minamisoma

Rob Roggema, Luke Middleton and Wanglin Yan

R. Roggema (✉)
Faculty of Design, Architecture and Building, University of Technology Sydney, Ultimo, NSW, Australia
e-mail: rob@cittaideale.eu; rob.roggema@uts.edu.au

L. Middleton
EME, Melbourne, Australia
e-mail: luke@emedesign.com

W. Yan
Faculty of Environment and Information Studies, Keio University, Fujisawa, Kanagawa, Japan
e-mail: yan@sfc.keio.ac.jp

R. Roggema and W. Yan (eds.), *Tsunami and Fukushima Disaster: Design for Reconstruction*, DOI 10.1007/978-3-319-56742-6_8

IMPACT OF NUCLEAR DISASTER

In the hills of the Fukushima region the radiation fall out was significant.

1. Geiger counter reading shows radiation levels remain elevated.
2. Contaminated soil is being relocated and buried under a track.
3. Removal of contaminated soil.

IMPACT OF NUCLEAR DISASTER

At the time of the tsunami and subsequent power plant melt down, the train stations were immediately evacuated.

1. Disused train-line enveloped with nature.
2. Bicycles parked at the local train station remain.
3. Abandoned train station platforms taken over by vegetation.

IMPACT OF NUCLEAR DISASTER

Many places in and around Minamisoma remain dormant since the disaster.

EXPLORING THE AREA

Meetings with local experts and community.

1. Presentation of happenings during and after the tsunami.
2. Visit to local horse track - this track has significant historical meaning.
3. Test case rice paddies analyzing the effect of local water supply on rice production.

HISTORIC SHRINES

An important part of our research included the ancient culture and rituals, this shrine visit is one example.

BE INFORMED

HORSE RACE

NEW BUSINESS OPPORTUNITIES

Working with local stakeholders and experts we investigated and visited a number of post-tsunami initiatives.

1. Solar farm where solar panels generate electricity whilst providing partial shade to food production.
2. Solar production and food production are being monitored with and logged with innovative computer systems.
3. Adaption - local farmer utilizing water treatment system for productive greenhouse growing.

SOLAR SHARING

IMPACT OF TSUNAMI

1. Clean up and reconstruction in progress after the old sea walls had been destroyed.
2. Twisted wreckage of previous infrastructure.
3. Evidence of the height of the tsunami can be seen by the destruction of the overpass bridge.
4. Low lying lands will continue to remain vulnerable.

THE IDEAL LANDSCAPE

Traditional rural Japanese farming communities locate their homes in the higher parts of the valley. Productive land with no permanent infrastructure can remain within the flood susceptible zones.

OTA RIVER

ENJOYING JAPANESE FOOD TOGETHER

LOCAL LIVING AND EATING

Sleep in a traditional Japanese hotel

Eat a traditional Japanese barbeque.

ENJOYING JAPANESE FOOD TOGETHER

TRADITIONAL HORSE RACING

For many centuries the town there has been an important horse racing event based on an ancient tradition in the region.

30 YEARS IN THE PAST

Community consultation event involving young and old. Groups were asked to contribute their knowledge and memories of the past. This exercise uncovers the collective remembrance of the area, their home.

PRESENTING

A summary of all the findings is presented by one of the group members.

30 YEARS INTO THE FUTURE

The participants were asked to put forward their visions for the region.

DISCUSSING THE FUTURE

At the end of the exercise a summary of all the findings is presented by one of the group members.

LOCAL SUPERMARKET: VENUE FOR DESIGN TEAM

A pop-up office was set up in the rear part of the local supermarket.

ANALYSING 30-30 RESULTS

Using the maps and the annotated post-it notes, the group set about two analyse the results, and find links and patterns within.

SKETCHING

In order to find a number of alternatives that could respond to the analysis, we formed two groups. Each group worked on design response ideas that were tested on large scale maps of the area.

This enabled us to develop two very different concepts based on the research and exploratory activities.

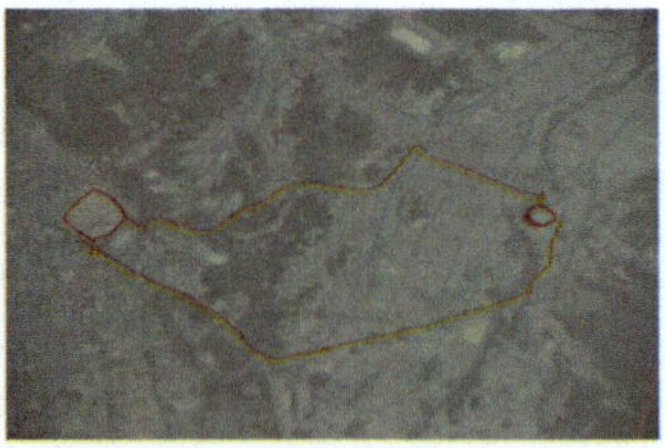

EXPLAINING

The early concepts were presented and explained to the local stakeholders.

PLASTICINE BUILDING

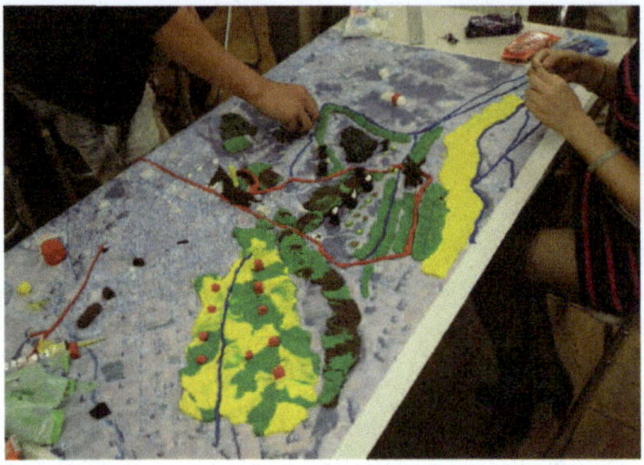

Taking the two dimensional map concepts and creating a 3d plasticine model.

Working in plasticine helps to break down the barriers between professionals and local stakeholders, whilst also enabling participants to use adaptive symbols in order to represent ideas.

RETHINK, REBORN, RETURN

Rob Roggema, Luke Middleton and Wanglin Yan with concept map

HORSE FESTIVAL ROUTE

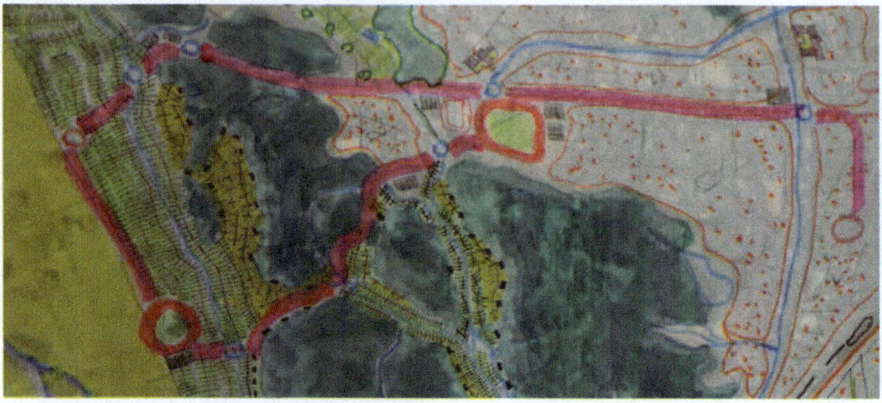

2d map and 3d plasticine model of concept masterplan. The red indicates the traditional route of the horse festival. In the masterplan this route becomes activated all year with new nodes.

HORSE KEEPING AREA

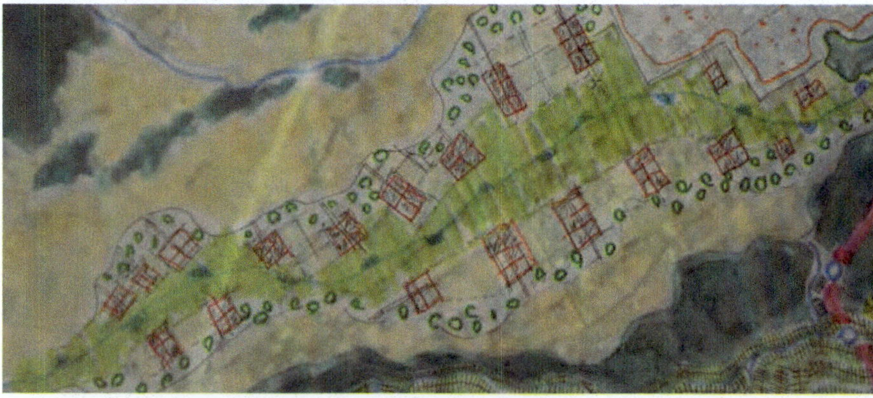

As part of strengthening and rekindling the importance of horse-related activities, the plan incorporates areas for horse accommodation.

COAST

ALGAE FIELDS

The low lying areas that were deeply flooded by the tsunami are re-imagined as algae fields that can be harvested to provide green energy.

This works in conjunction with a perforated dam/sea wall that enables this low line land to remain flooded.

The low lying land adjacent to the Ota river, which has been affected by radiation fall-out is allocated to become canola fields.

During our analysis it was discovered that the local farmers and scientists had developed this crop and which does not transfer any radiation to the harvested produce.

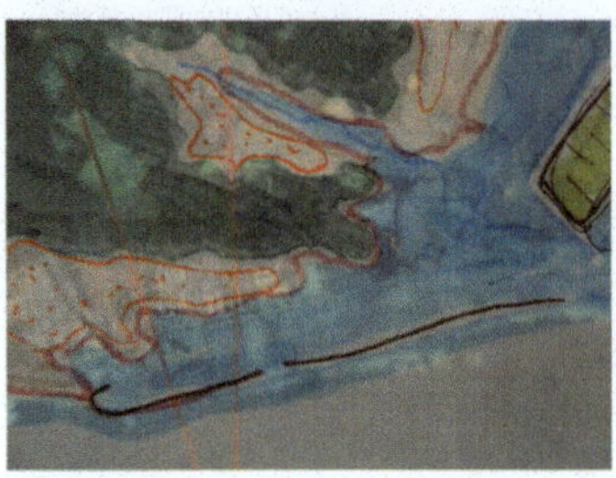

PERFORATED DAM

LAND-USE MOSAIC

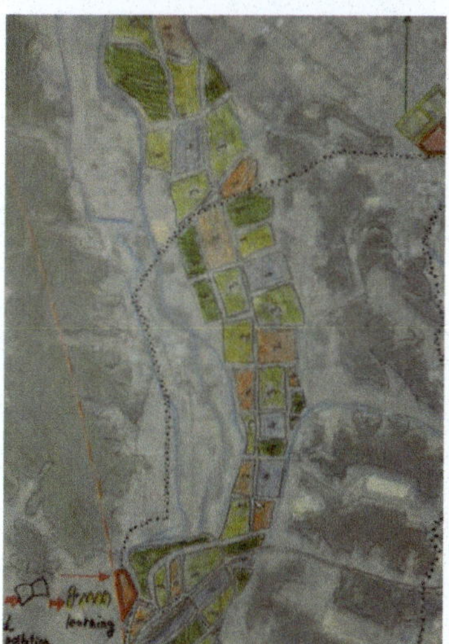

Group 2 had an alternative approach - to divide many of the lands into a mosaic garden.

This included solar farm, vegetable gardens, forest, Japanese gardens and rice paddies, which were located along the Ota river.

GROUP 2

HORSE RIDING SCHOOL

SURFING
BEACH PARK

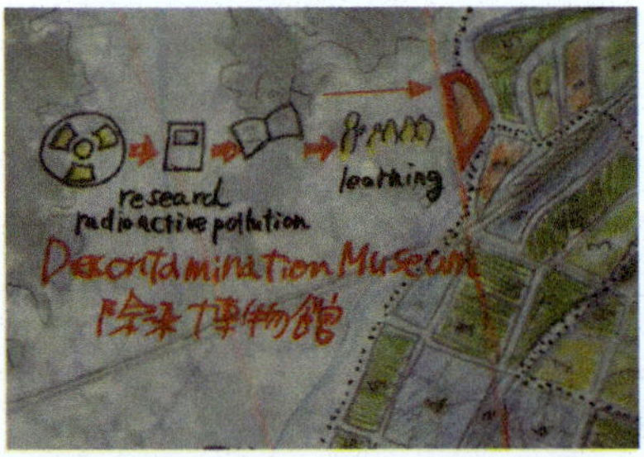

DECONTAMINATION
MUSEUM

EXHIBITION OF DESIGN RESULTS

Local stakeholders were invited to view the concept map and plasticine model at an evening event.

PRESENTATION OF DESIGN RESULTS

Each group did a short presentation to provide an explanation of their concept and how it was derived.

PLASTICINE MODELLING

The stakeholders were invited to create their own plasticine model

PLASTICINE MODELLING

The activity was embraced by young and old with lots of enthusiasm and laughs.

PLASTICINE MODELLING

It was evident that the Japanese craft was inherit in many of the participants.

PRESENTING WORKSHOP RESULTS

Each group of stakeholders presented the ideas and concepts within their plasticine model of the area.

GROUP PICTURE

Chapter 9
Visualisation of Kesennuma

Rob Roggema, Luke Middleton and Wanglin Yan

R. Roggema (✉)
Faculty of Design, Architecture and Building, University of Technology Sydney, Ultimo, NSW, Australia
e-mail: rob@cittaideale.eu; rob.roggema@uts.edu.au

L. Middleton
EME, Melbourne, Australia
e-mail: luke@emedesign.com

W. Yan
Faculty of Environment and Information Studies, Keio University, Fujisawa, Kanagawa, Japan
e-mail: yan@sfc.keio.ac.jp

R. Roggema and W. Yan (eds.), *Tsunami and Fukushima Disaster: Design for Reconstruction*, DOI 10.1007/978-3-319-56742-6_9

ENUNKAN

Traditional Japanese tea house with residence adjacent.

View of tea house from Japanese garden

View of tea house from Japanese garden

OVERVIEW

Views of tsunami devastation area, currently under clean up and reconstruction

FISHING HARBOUR

The area once supported a thriving fishing community.

Remanents of the sand bagging - an in vain attempt to stop the tsunami.

FOUNDATIONS

Foundation stones of buildings that once stood in this place swept away and destroyed by the tsunami.

BUILDING A NEW LEVEE

Current works in the construction of a new levee to one side of the river mouth.

IMPACT OF TSUNAMI ON OLD SEAWALL

Massive concrete seawalls failed to hold back the immense power of the tsunami.

IMPACT OF TSUNAMI

Road barrier mangled by the force of the tsunami and objects that were swept away in its path.

IMPACT OF TSUNAMI

New seawalls under construction by the government have mixed support amongst local residence.

NEW SEAWALL

Many peoples remain in temporary housing installed immediately after the tsunami.

TEMPORARY HOUSING

BEING INFORMED

Discussions at the temple with local experts and stakeholders in order to understand what has been done since the tsunami. And importantly what they see as the opportunities for the future of the area.

Discussions at temporary housing project with manager

Discussions at the tea house

SHRINE ON THE 'SAFE-HAVEN' CAPE

Elevated sacred places become safe-haven.

View of ridge with shrine amongst vegetation. These spaces were not affected by the tsunami.

TEMPLE ON THE HILL WITH MONUMENT FOR VICTIMS

View of temple

View of Monument for Victims.

ALONG OIRAN RIVER

Tracing the path of the Oiran river.

Walking with local farmer to small shrine site. Looking for the Cat Stone.

SHRINES & SYMBOLS

CAT STONE

MIRROR
STONE

SHRINE OF FERTILITY

LANDSCAPE OF OIRAN RIVER

Rice paddy with home nestle into the edge of the forest.

Terraced rice paddies in the lower land adjacent to the water way.

Remanent bamboo forest.

OIRAN VALLEY

The low lying topography was severely affected by the water inundation, taking out major roads and destroying both man-made and natural landscape.

WASHED OUT CEMETERY

NEW MAIN ROAD

FISH

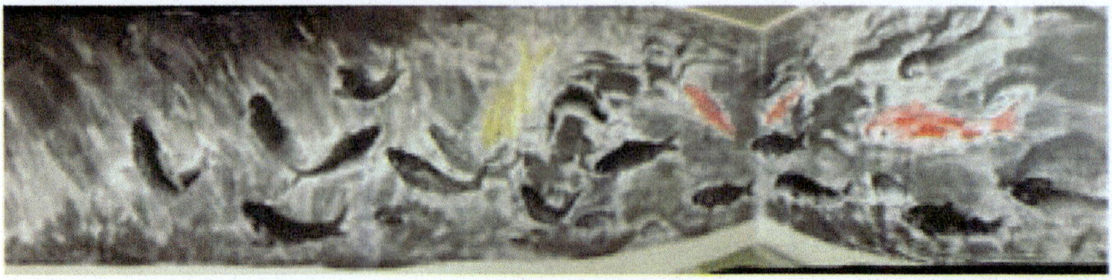

Fishing was a large part of the culture dating back many centuries. However, with the advent of the tsunami and aging population keeping this important industry and tradition alive is proving challenging.

FOOD

Sharing time with local friends while enjoying magnificent local food.

SUSHIMI

BARBEQUE

BENTO BOX

FISH & FOOD

30 YEARS INTO THE PAST

Mapping workshop with local stakeholders to uncover the memories and knowledge of the past 30 years.

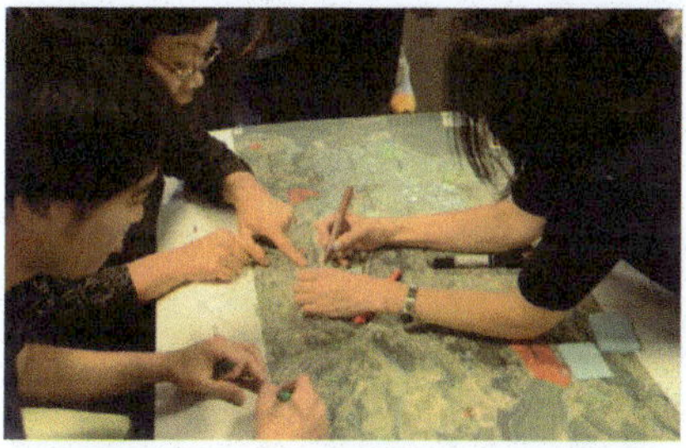

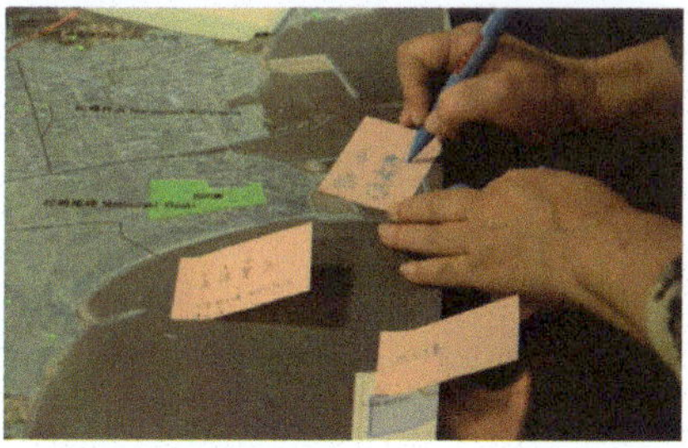

THINKERS

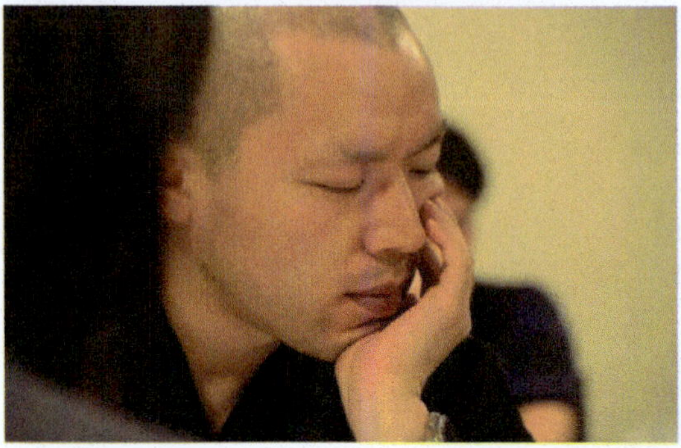

Stakeholders engaging in mapping workshop.

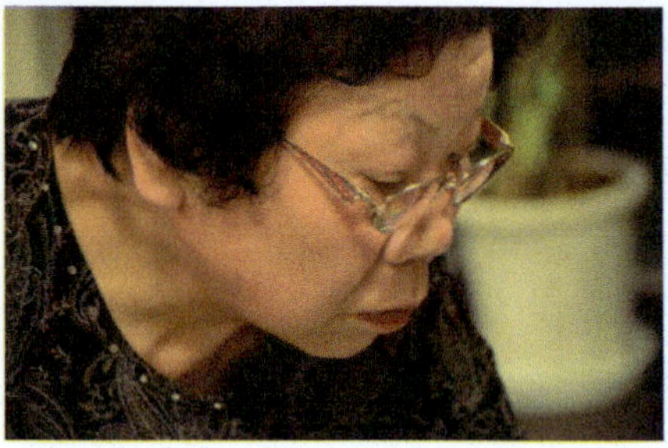

30 YEARS INTO THE FUTURE

Stakeholders invited to share their future visions for the area to uncover their dreams and aspirations.

PRESENTING RESULTS

The workshops were completed with a presentation by each stakeholder group.

ANALYSING & SKETCHING

Setting about to analyse the inputs from stakeholders.

Interpretation of results via sketching on large scale maps to evolve patterns and ideas for the masterplan.

DESIGN DELIBERATIONS

Working through design possibilities and options.

Reviewing the design and plasticine model.

REMEMBER, RECONNECT, REFORM

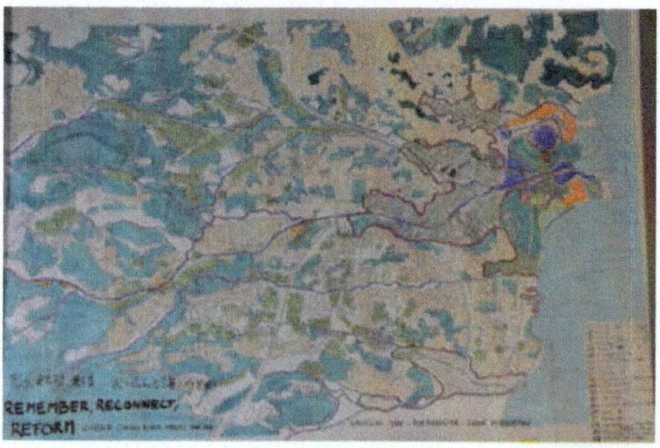

Masterplan map.

Masterplan plasticine model.

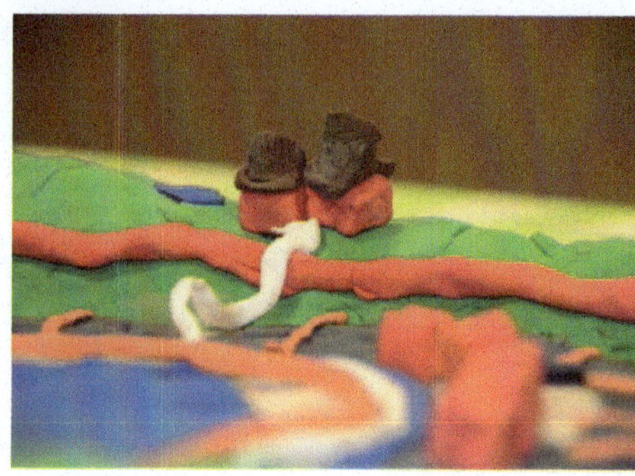

CHERRY LINE IN DESIGN & MODEL

A continuous line of cherry trees would be planted to delineate and remember the high line of the tsunami.

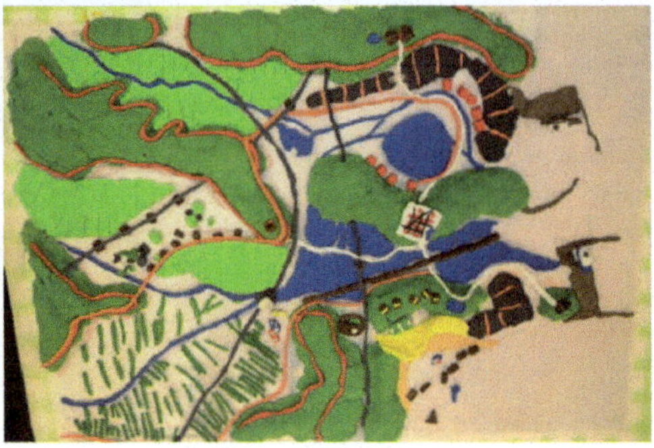

MONUMENT SQUARE

The masterplan envisage an intervention of large raised landscapes that would remain safe and resilient during a tsunami event.

Following the ancient planning tradition a monument square is constructed on top of the elevated spaces.

This square adopts and is inspired by the footprint of the fishing residence zone that was demolished during the tsunami.

Plasticine model of the raised lands and Monument Square.

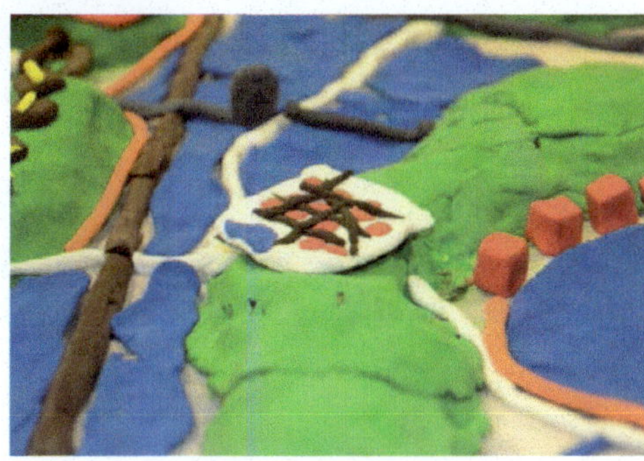

SPORTS & BEACH AMPHITHEATRE

Other parts of the raised landscapes are set aside for sporting infrastructure including amphitheatre, baseball park, tennis court, swimming pools.

The existing beach is partially protected by a permeable seawall which is designed to break down the energy of a tsunami.

OTHER POINTS OF INTEREST

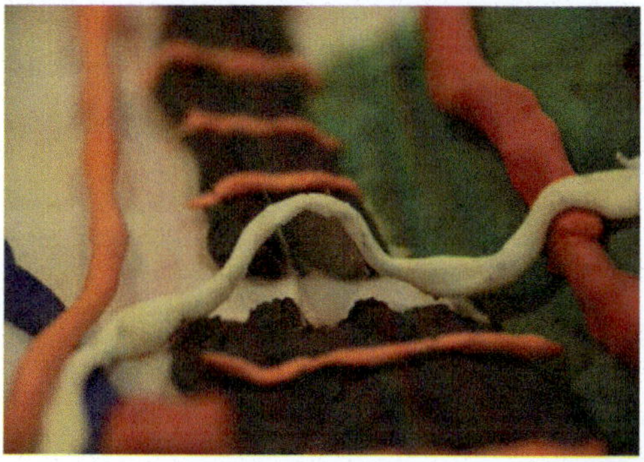

Subsoil foundation watchpoint

Proud presenters

PLASTICINE WORK

Stakeholders invited to design workshop.

Working with plasticine they created their own vision of the future on the map of the area.

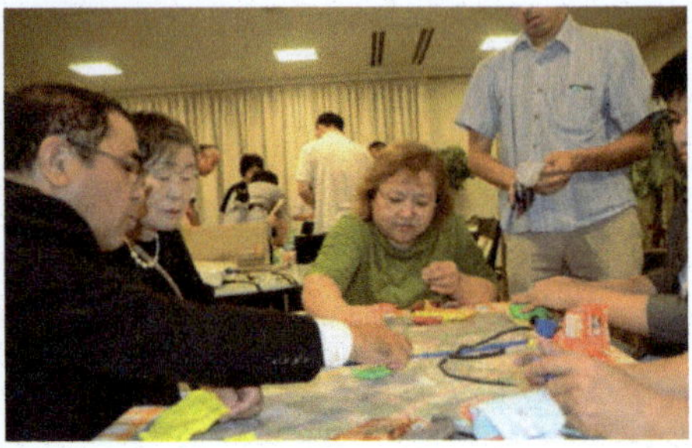

TACTILE

The plasticine is a medium that breaks down the barriers between professional and stakeholders.

It is tactile, adaptive and enables people to convey their ideas symbolically within a spacial context of the map.

CO-CREATION

Groups worked together finding harmony amongst their individual ideas to create their overall concept.

PRESENTATION OF GROUP CONCEPTS

Marine Park

Yaught Harbour

Transition from fishing harbour to tourist harbour.

Sea/aqua Museum

Guest rest-house

Accommodation, seminar house, close to garden and regional library,

Koibito (lovers) lane, dating laneway

Cherry lane, on the levee, cycling course

East Sunshine Festival in Old house/garden

Three storey factory

Experience centre,

Oyster processing,

Experience, products, learning fishing,

Watch fireworks on island from top floor,

Three storeys required to have a look over the levee

Hotel, pool

Squid Factory

Sports facilities, soccer, baseball, parking, shopping, restaurant mall, athletics track, rent a cycle

Recover local train along the coast, or bus rapid transport (BRT)

Japanese garden

Deer, monkery, North hill

Inland: BBQ camp site,
Folklore story area,
stepping stones,
locations along the river

Fox and the cat story

PRESENTATION OF GROUP CONCEPTS

Up the river: Folklore stories, folklore road along the river

Small rooms, each have a different story about fox or cat

Cat saved a life

Hydro pump in the river

Gold mining area, government office for taxes

Two big roads (big recreation route) to the mountain

Evacuation road, connect to main road

Squid Factory

Skate park, curling park

Beach

Japanese garden

Warm water pool

Sports park, cultural centre, event park, fishing

Yaughts

Temple, Buddhist Memorial Park, Aqua Museum

Fishing

Parade of the Daimio, people, generals, kings of the land, collect people in traditional clothes, from shrine to mountain and schools. Disease and plague in the area brought strong men together, stopped the disease. Took place in 2000 for the last time.

Ghost Story, Devil.

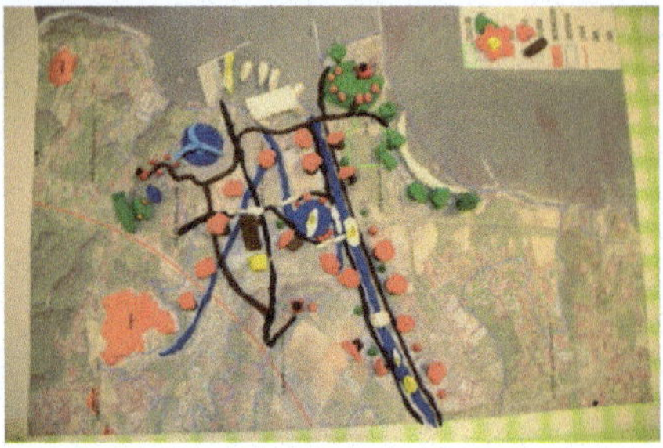

PRESENTATION OF GROUP CONCEPTS

Big natural park

Marine park, dome city, sports park (tennis, football, baseball, multi-purpose square, spring/fountain square, lantern, kites), promenade to the shrine for souvenir & local food selling, yaught - harbour ferry.

Lovers dating beach, love square

Japanese garden, old garden/house

Cafe

IC, hotel, shopping mall

Cycling course along the river with colourful plantation in different seasons

Memorial centre for disasters

Local food centre/market/restaurants

Learning centre for fishing skills, sea farming

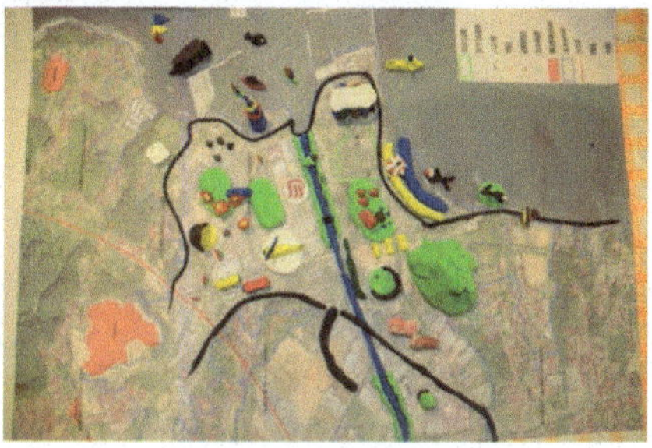

FINAL RESULTS

Plasticine models on display.

TEAM PICTURE

Chapter 10
Conclusion, Recommendations and Outlook

Wanglin Yan and Rob Roggema

In this book the design charrettes, which were held in Minamisoma and Kesennuma, are presented as examples of resilience planning in complex circumstances. In Chap. 1 five stages of planning for recovery have been introduced. Recovery, reparation and rebuilding are processes, which should start immediately after the disaster took place. In various dimensions, such as the demand for immediate aid, but also the huge media attention, the urgency to take on recovery, reparation and rebuilding processes is clear. Resettlement and redesign receive less attention, if at all. The two exemplary design processes presented in this book make the benefits clear of a focus on redesigning an area that suffers from the impacts of a natural disaster. It illustrates there is as much urgency to redesign an area as there is for immediate recovery. The main difference between the two is the long and short terms each play their role. Without immediate care the area will never recover, but without a fundamental redesign the area will be trapped in the same vulnerabilities for even worse disasters in the future. Therefore it is important to keep an eye on the long term of planning, rehabilitation and sustainability in a disaster struck area. And, even more so, what is valid for disaster prone areas, is also recommended for areas that do normally not suffer from a disaster.

W. Yan (✉)
Faculty of Environment and Information Studies, Keio University,
Fujisawa, Kanagawa, Japan
e-mail: yan@sfc.keio.ac.jp

R. Roggema
Faculty of Design, Architecture and Building, University of Technology Sydney,
Ultimo, NSW, Australia
e-mail: rob.roggema@uts.edu.au

R. Roggema and W. Yan (eds.), *Tsunami and Fukushima Disaster: Design for Reconstruction*, DOI 10.1007/978-3-319-56742-6_10

10.1 Toward a Co-creative Reconstruction and Community Planning

Reconstruction work since the 2011 disaster is a year or two behind schedule. Besides the cases covered in this book, reconstruction has not yet begun in some areas, in other parts of Fukushima and outside of municipalities in Miyagi Prefecture. Nevertheless, despite the delays, plans can indeed be developed through the coordination of intermediate support organisations and with citizen participation. As shown in the two examples in this book, the process of recovery from a disaster of this magnitude is complex and not something that can be accomplished by someone working alone. In this case, four elements came together: intermediate support organizations, the resident community, spatial planning by experts, and reconstruction projects, with the ambition to turn it into a co-creative reconstruction and community planning with the aim of reconstruction.

Under this approach, residents gained awareness of the issues intermediate support organisations noted this and arranged external resources to create the space for co-creation through actions that included this design workshop. There was discussion and mutual learning, and a reconstruction vision was then formed, making use of the local, natural characteristics and cultural resources. A spatial plan known as a landscape design was then put together and transformed into concrete form during the process. The vision for reconstruction is like a banner or flag used to guide the troops in ancient battles, and is a concrete expression of the target region for a higher-level reconstruction plan. For this to take shape, it is necessary to incorporate in a comprehensive way the policy directions, local wishes and regional physical and social resources as well as constraints. For that to work, the information gathering and coordination capacity of intermediate support organisations is important. Residents may not have expertise relating to individual projects, but residents are the ones who know their region the best. A spatial plan incorporates the wishes and intentions of residents and then fits a variety of reconstruction projects into a spatial format, while enriching the physical and social resources that are unique to the region. The resulting projects represent both "hard" (physical) and "soft" (institutional) actions that can turn a reconstruction vision into reality.

10.2 The Role of the Design Charrette Workshop

Co-creative reconstruction and community planning is a social learning process. The design charrette is a platform to support that process. Looking at the cases of Minamisoma City and Kesennuma City, this learning process was conducted through the iterative processes composed of:

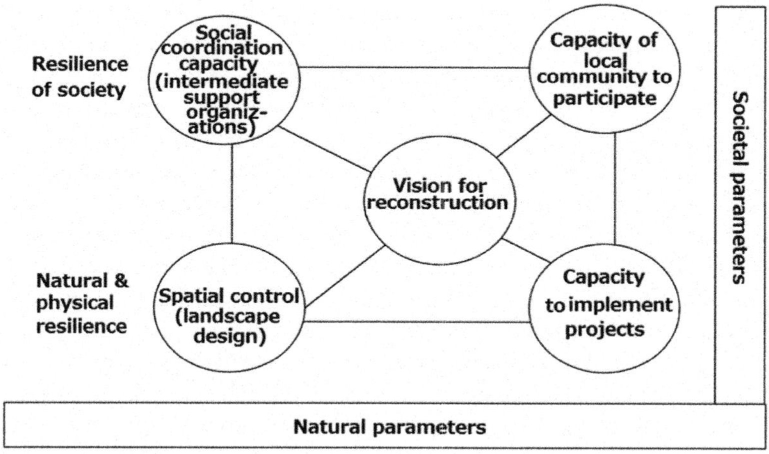

Fig. 10.1 Co-creative reconstruction and community planning, and building resilience

1. The collection of local information
2. The creation of designs by experts
3. Presentations
4. Citizen participation design workshops.

The design proposal the experts produced aims to offer a sustainable development model rooted in the local natural, cultural, social, and economic resources (Fig. 10.1). There are clear differences here compared to conventional conceptual reconstruction plans. The residents design workshop created learning benefits, because it took place after a presentation by experts and a three-dimensional design demonstration.

Residents have a strong interest in the reconstruction of their town, and are full of desire to participate. However, working alone, residents are not fully able to bring things together organizationally and spatially. This is where intermediate support organisations and experts have an important role to play.

10.3 Building Resilience

Looking at things in terms of "resilience" provides synergistic ways of seeing problems, in terms of time (normal times, times of disaster, times of reconstruction, etc.), in terms of space (project, district, overall town, etc.), and in terms of stakeholders (individual, organisation, government, etc.). This concept of resilience may seem difficult, but putting it into practice makes it very concrete. For example, there is the engineering concept of raising the seawall for protection from a tsunami, but there is also the "soft" idea that the escape routes should be improved to permit rapid evacuation. In addition, even if everyone is supposed to evacuate, there is still uncertainty or concern about the elderly and disabled. Resilience can be viewed in a specific way. Or resilience can be viewed as to build a lively community over the long term (Walker et al. 2004). Disaster prevention, disaster reduction, reconstruction, and recovery projects are time-limited, but their cumulative effect is to create the capacity to prepare for the next disaster or change, and to increase resilience. One approach to undertake this is presented in this book, as the framework of co-creative reconstruction and community planning.

The reconstruction vision: It is necessary to take the reconstruction targets announced immediately after the disaster and customise them to fit the region where reconstruction is to occur. This is not something to be decided top down, but rather, bottom up. Repeated consultation is needed with citizens, experts, and government agencies, and it should be done comprehensively, considering things such as local natural conditions, social conditions, the policies of high-level plans, social trends, and residents' desires. In the post-war society with a growing economy, the things valued by society were relatively straightforward, and it was relatively easy for governments to respond. In today's mature Japan, however, the future of the economy and society is not so clear. Creative ingenuity is needed case by case, looking at what a specific type of community aims for. In the case of Minamisoma, mentioned as an example in this book, the city is aiming for reconstruction as a town that uses renewable energy, and it has special local characteristics of samurai-style traditional wild horse racing and history of a philosophy known locally as "Goshiho" (closely connected with today's concepts of local sustainability and self-sufficiency, from Sontoku Ninomiya, a prominent 19th-century Japanese agricultural leader and philosopher who lived 1787–1856). With a focus on this orientation, it was decided to do a landscape design with the vision of a community that is self-sufficient in terms of energy. In the case of Matsuiwa and Omose, (Kesennuma) the cities aim for "slow and smart" concepts, and the reconstruction vision is to develop a centre of synergistic cultural creation, because the target area was a decent distance from town centre, and there is active interplay between a unique culture and landscape resources, the sea and mountains.

In order to make the reconstruction vision feasible for implementation, it is necessary to approach things from both physical and social dimensions. It is essential to have active institutional and resident participation to make reconstruction objectives spatially concrete on the physical dimension, and to coordinate between diverse stakeholders on the social dimension. When it comes to developing plans, experts have expertise, but if they have no capacity to adjust and make local refinements, they cannot gather information from the frontlines hence the local residents, and they cannot create a plan with its feet solidly "on the ground." A project transforms into concrete form over time in between the higher-level plans and local reconstruction plans, but without the involvement of residents and experts, it ends up simply as an "inorganic" public works project. In the Matsuiwa and Omose district in Kesennuma, an enormous concrete dike is appearing in Omose, but it creates a sense of crisis among residents.

In co-creative reconstruction and community planning, residents are the main actors. Reconstruction projects are just one business area for the contractors and businesses involved, but for the region and its residents, that construction could dramatically change the community. To expand the positive effects and minimise the negative ones, residents need to be brought into the process at an early as possible stage, being the "local experts." In this endeavour, "participation" does not just mean receiving repeated explanations from the project implementers. It means that diverse stakeholders need to get together and consider things comprehensively. By this approach, it will be like the implementers and projects will absorb nutrients from the land, and be able to realize co-creative reconstruction community planning.

The need for and importance of intermediate support organisations has been described in many case studies. There are many kinds, including non-profit organisations, registered associations, universities, and consortiums, etc. In Minamisoma there was the Eco-Ene Minamisoma Research Laboratory, and in Kesennuma, the Association to Create a Liveable Kesennuma City. Each organization arose from a different background, but their common point is fulfilling a role as go-between between government agencies and the residents. Both organisations are knowledgeable about the local situation, and are proficient at organising and coordinating. They were also good at exchanging information both internally and externally, and motivated to achieve reconstruction and community planning. By making use of intermediate organisations like these, it is possible to bring together a variety of resources in one place.

A reconstruction project is something that puts a vision for reconstruction into concrete form, with a "hard" (physical) spatial plan and "soft" projects. National and local governments often undertake large-scale projects, but it is preferable to bring this down to the local level at an early stage of the process, and get specific in a unified way with a vision for reconstruction.

It is expected that co-creative reconstruction and community planning increase resilience through this approach, by learning from pre-disaster development and creating a vision that includes local self-reliance and sustainability as important characteristics.

10.4 Conclusion

Because of the complexity of the situation after the Great East Japan Earthquake and the uncertainty of the political and economic environment, reconstruction of this area is a long-term strategic challenge. The issues being confronted in reconstruction are issues that Japanese society has faced for many years, but the disaster has brought their timing forward by fifteen years. Just as Japanese society went through dramatic changes after the war and Japan joined the ranks of developed countries, can the country utilize this post-disaster period to put Japanese society on the path of sustainable development (Mikuriya 2016)? But differences in the situation must be acknowledged. Post-war is post-war and post-disaster is post-disaster. It is difficult to recreate a model that succeeded in the past. For the problems of today, the post-war model can also be a liability. Today's socioeconomic system is built upon large cities, large industries, the large movement of material, and large markets. It has left little space for regional development. For reconstruction, something that aims to using an old model has no future. It is necessary to find a new paradigm using the disaster as an opportunity. For that, everyone has a role to play, and collaboration is essential, among citizens, the government and government agencies, academics, and businesses. The residents include landowners, but they are also experts about local history, culture and resources. They need to regain their confidence and need to be provided opportunities to focus on their local area. Government agencies are leaders entrusted by citizens and they are managers of assets, and have major responsibilities relating to the future of the entire community and local areas. Governments decide on reconstruction strategies, and prepare or create a flexible environment to put plans into action. Academics can apply their broad knowledge to grassroots problems, and research ways to resolve and address specific and nuanced challenges. Companies need to develop business approaches that can make business feasible on the small and medium scale in less urbanised parts of the country.

This book illustrated the potential for creating a reconstruction vision through close linkages of these elements, by putting co-creative reconstruction and community planning into practice. This approach is based on experience gained in practice in Minamisoma and Kesennuma, but reconstruction anywhere faces some similar challenges and there is a certain degree of universality in the concepts themselves. The word "co-creation" is often used in community planning. This book discussed, within the specific frame of reconstruction, how to consolidate ideas and point the long-term vision, time-limited projects, residents' perspectives, and expert knowledge in one direction. It must be noted, however, that both the examples covered in this paper are still in the midst of reconstruction, and currently only at the point of a landscape design being presented. Both examples represent voluntary activities by residents, not something imposed by government agencies. The aim of co-creative reconstruction and community planning is to receive a variety of stimuli through a co-creative approach, trigger action by more people, and to align the "vectors" into a vision for reconstruction. Fortunately, in Minamisoma and Kesennuma, intermediate support organisations continue to be active, and are tackling reconstruction based on the spirit of co-creation. For example, in Minamisoma, a seaweed processing plant is being built in the area that was inundated by the tsunami. In Kesennuma a survey project has begun in the Matsuiwa and Omose district, and residents started a citizens' garden project. If activities like these can continue to build momentum and grow into a larger movement, the vision for reconstruction can probably be achieved.

The lessons learned in the design processes are varied. The first conclusion derived from these processes is that the people, due to their direct involvement in co-creating the design, will be long-term connected to the area and this leads to a higher commitment to help with projects or support fellow citizens, and take responsibility for realization of a resilient future. Secondly, the process of the design charrette with its tactile methods, which visualise the future, has a bonding effect on people of the area. The third conclusion regards the role of the planner or designer. He or she should take a facilitating role in the process, and use their skills to support the design process of the entire group. The variety of participants within the design group is important. There is also a benefit in working intercultural. Insight from people with different perspectives help to broaden the view, keep the solution space wide open and prevent people from being preoccupied. Conclusion five is that external eyes persuade. When people come from abroad they can design and bring some issues to the fore a local expert cannot do. This implies the subjects discussed and the ideas coming up are innovative and potentially ground breaking, exactly what is necessary when complex problems need to be discussed in a tense context, such as in the aftermath of a disaster.

As far as the content is concerned, in both exemplary plans there is a strong focus on the cultural and historic component of the area. The horse related activities in Minamisoma and the storytelling in Kesennuma have proven to be crucial elements in designing a resilient future. Besides this, both plans resemble a powerful yet smart and sensitive response to safety. In both plans is also a strong attention for resilience and sustainability. The focus on solutions that give space to multiple uses, a coastal zone that is dissipative and welcomes the sea, and a strong emphasis on becoming self-sufficient in its energy-use, water use and food consumption, is in each of the designs evident.

The planning approach presented in this book could be beneficial in the aftermath of a disaster but also in preparation for a potential disaster. Redesigning an area in order to make it less vulnerable for any disaster or other problem in the future is good for the region, its inhabitants and the government. Also it is cost-effective to prepare a region for future change instead of responding to unforeseen events and repair the damage afterwards.

The main difficulty of this approach is to make clear what the benefit of the approach is before it is applied. It is easier to showcase strong, but expensive measures for dealing with the impacts of a disaster. It shows strength and vigour of decision-making officials and politicians, gives inhabitants an immediate image of action and response. However, this power used in the beginning might fail on the long term. As soon as people are confronted with the effects of these vigorous measures they might turn their back on them and confrontation looms. If instead the inhabitants are co-creators of future plans this will not happen easily, as the people are committed. The exact and clear outcomes of such a process are however uncertain and invisible at the beginning, which makes it hard to explain and gain support.

The main lesson to learn for other areas in the area where the tsunami hit in preparing for a disaster, is to have strategies in place for each of the five stages of recovery planning: recover, repair, rebuild, resettle and redesign. In this context, as the exemplary plans illustrate, resettle and redesign strategies can be a permanent effort of a municipality, while the other strategies should only be used when a disaster actually has happened. Especially for vulnerable areas it is essential to formulate these five strategies for the times to come. Once a disaster happens it is often too late to think of ways to respond to the occurrence.

Another lesson learnt is the use of design charrettes in the process of involving residents in the design process. As mentioned above this will commit these people to the area, makes them proud and gives them the power to act. For every municipality it is beneficial to use this approach to create a shared vision for future resilience. This method is also very useful when federal governments would like to discuss the future with local people.

A third lesson learnt is that when people from outside the community are brought in the deign process, innovative plans are easier been developed, because these people, who preferably come from abroad, have the capability to think outside the beaten tracks and usual pathway of solutions a local community often has. The beauty and uniqueness of the place is often not very clearly seen by the residents, but will be highlighted when people from outside are brought in the discussion.

Other countries can benefit from this model as much as in the Japanese situation. The method of a design-driven process gives the opportunity to tackle complex problems, to involve people from different backgrounds, to involve local residents and to think of solutions that have never been thought of before. The future is inherently uncertain and therefore an approach should be installed how a community would like to future to be. This way the future can be partly directed and the people living in vulnerable places are not left to their fait.

10.5 Recommendations

The design methodology used in Minamisoma and Kesennuma showed how citizens in a disaster struck area can be involved in a design process and produce a regional spatial vision to minimise vulnerability, increase resilience and sustainability, and creates a landscape in which people want to live, work, produce and leisure.

The first recommendations is therefore to use this methodology in other regions in the tsunami area in Japan, and where applicable in other vulnerable areas in Japan and beyond.

Secondly, it is advised to involve experts from outside the area, preferably from outside Japan, in these kinds of design processes. External opinions thrive the debate in local communities towards innovative and unusual solutions for pressing problems.

In the third place, it is recommended that municipalities formulate a framework in which the strategies for each of the five stages of recovery planning are used. This implies formulating a policy in which anticipative plans for recovery, repair and rebuilding are put in place, even before a disaster has occurred. Moreover, regular planning and design should focus on resettlement, e.g. increasing resilience in the area, and redesign the area in a way it is able to withstand disasters and function in a sustainable way.

The fourth recommendation is to form an international group of experts in community driven design for disaster recovery planning, which could be of help if communities would like to set up the recovery framework, or if they need help after they suffer from a disaster.

The next advise is to make the community driven design methodology, as described for the exemplary projects in Minamisoma and Kesennuma and the framework of five stages of recovery planning, part of the Sendai framework.

Recommendation six is to implement the outcomes of the design processes through:

(1) Realisation of catalyst projects, which are the icons of the redesign and are visible to the inhabitants. These catalyst projects should start the process of realising more projects.
(2) Create a group of people (entrepreneurs, decision-makers, citizens and stakeholders) and make them responsible for the realisation of projects.
(3) Create an investment fund that can be allocated to the projects derived from the redesign.
(4) Create a board of supporters, supervising the process of realisation, preferably people that have participated in the design charrettes.

Finally, it is advised to spread the new knowledge that was gained during the design processes through dissemination of the results (publications), a yearly seminar or conference, an international student exchange, the set up of a Masters degree in community based disaster design, and/or a travelling design studio in which the participants make designs for actual disasters around the globe, no matter if t is an earthquake, tsunami, bushfire, cyclone or hurricane.

10.6 Outlook

Learning from the design processes carried out in Minaimisoma and Kesennuma a future outlook for planning and design in disaster struck areas should include the five stages of recovery planning as well as the design methodology used in the two example towns.

Wouldn't it be great if all vulnerable towns and cities around the globe use these methods.

Let's imagine ….

An imaginary town is vulnerable for flooding, an occasional earthquake, risk at bushfire in summer and landslides in autumn. In this town nobody is really afraid for these disasters. How come? Because this town is well prepared. Some people in this town started 15 years ago to bring a group of people together to think about the resilience of their town. This group consisted of active citizens, local leaders, entrepreneurs and few people with decision-making power. This group made a plan. The 'redesigned' the town and made it more resilient. The plan took into account the occasional disaster and people proposed measures that could respond to these events. They decided where the immediate aid can be found and is provided, and where which crucial infrastructure needs to be repaired. Just in case of a disaster. They've thought of how to rebuild parts of the town after an imaginary disaster, and implemented these anticipative plans in reality. What they came up with were the ingredients people would need just after they were confronted with a disaster: an anticipative, just in case, strategy. During the planning process, they started to spread the word, telling the story of designing this plan and sharing their ideas with their family, friends, at the sports-club, at work and school. Other became enthusiastic and started to think along. This snowball affect has led to involvement of the majority of residents, in one way or another, in the planning process: A Swarming way of advocating for this resilience city building movement. Once the ideas and plans for the time just after a disaster were elaborated, this paved the way for long-term thinking and created support for redesigning the fundamentals of the town, in order to make it less vulnerable on the long term. This redesign included shifting the settlement in a way the entire town became safer. Moving the places where people live to less vulnerable locations, away from flood- or bushfire-prone zones, at elevated locations and away from the impacts of earthquakes. This all is in place before an actual disaster eventually might happen. In case a disaster occurs, everyone in town knows where to find aid, where to move to and which parts need to be reconstructed first.

This resulted in a community which is coherent, prepared for all kinds of surprises, living in a town which can absorb unforeseen shocks, and is wealthy too, as the people chose for productive land-uses, such as local food production and energy generation, or the cleaning of water, from which everyone profits. The people know each other because they work together on realising the projects that were part of the regional design for the town.

Let's imagine this is not an imaginary town, but a real one, and imagine this is not the only one but a whole series of towns along the Pacific coast of Japan. Imagine it is a series of towns under threat of a disaster around the world.

Let's imagine.

References

Mikuriya. 2016. Thinking about Japan from Post-war to Post-disaster, Nihonkeizaisinbun, 10 Mar 2016. http://www.nikkei.com/article/DGXMZO97981080T00C16A3000000/. Last access date 24 Apr 2016 (in Japanese).

Walker, B., C.S. Holling, S.R. Carpenter, and A. Kinzig. 2004. Resilience, adaptability and transformability in social–ecological systems. *Ecology and Society* 9 (2): 5. http://www.ecologyandsociety.org/vol9/iss2/art5/ (online).

The manufacturer's authorised representative in the EU is Springer Nature Customer Service Centre GmbH, Europaplatz 3, 69115 Heidelberg, Germany. If you have any concerns regarding our products, please contact ProductSafety@springernature.com

Printed and bound by CPI Group (UK) Ltd, Croydon, CR0 4YY
07/01/2026
02031048-0006